THE FAR PLANETS

As the banded face of Jupiter looms over the horizon, an arcing fountain of sulfur dioxide *(lower right)* spurts a hundred miles into space from an active volcano on the moon Io.

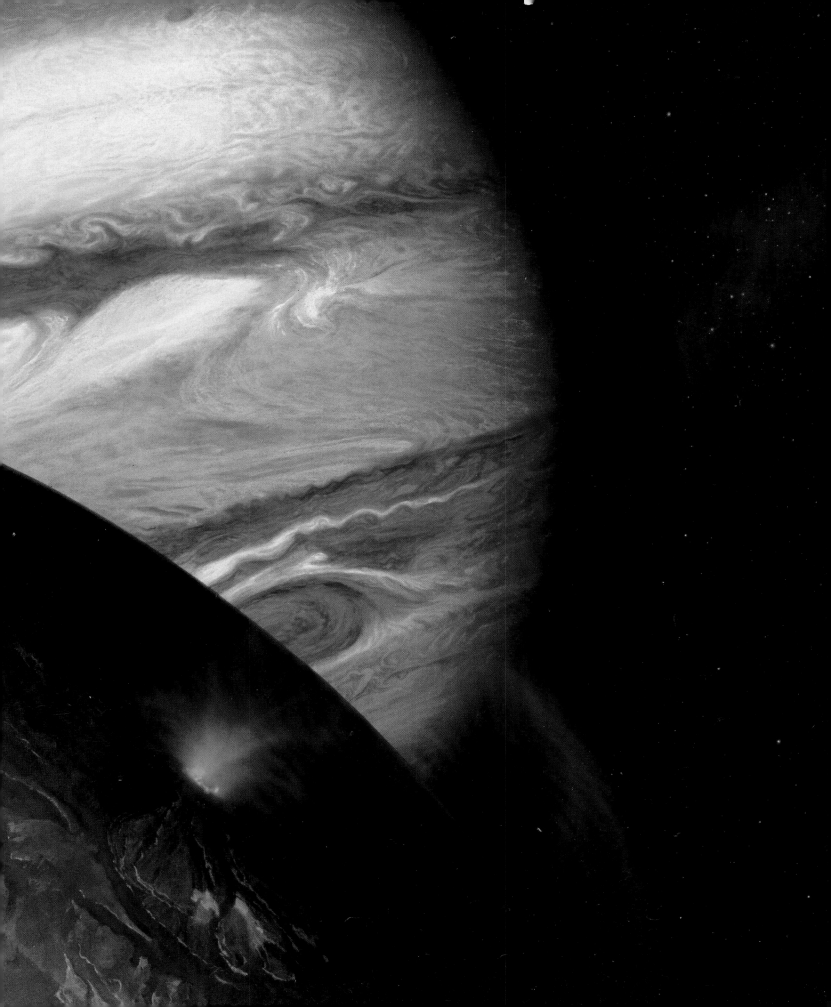

Haloed by its own refracted light, the rising
Sun enflames the atmosphere of the Saturnian
satellite Titan and turns the ringed planet
itself into a crescent moon.

The luminous blue globe of Uranus drifts out
from behind the rim of three-mile-high cliffs on
Miranda, a moon whose deeply scarred surface
suggests long-ago collisions.

TIME
LIFE ®

Other Publications:
THE TIME-LIFE GARDENER'S GUIDE
MYSTERIES OF THE UNKNOWN
TIME FRAME
FIX IT YOURSELF
FITNESS, HEALTH & NUTRITION
SUCCESSFUL PARENTING
HEALTHY HOME COOKING
UNDERSTANDING COMPUTERS
LIBRARY OF NATIONS
THE ENCHANTED WORLD
THE KODAK LIBRARY OF CREATIVE PHOTOGRAPHY
GREAT MEALS IN MINUTES
THE CIVIL WAR
PLANET EARTH
COLLECTOR'S LIBRARY OF THE CIVIL WAR
THE EPIC OF FLIGHT
THE GOOD COOK
WORLD WAR II
HOME REPAIR AND IMPROVEMENT
THE OLD WEST

This volume is one of a series that
examines the universe in all its aspects,
from its beginnings in the Big Bang to the
promise of space exploration.

VOYAGE THROUGH THE UNIVERSE

THE FAR PLANETS

BY THE EDITORS OF TIME-LIFE BOOKS
ALEXANDRIA, VIRGINIA

CONTENTS

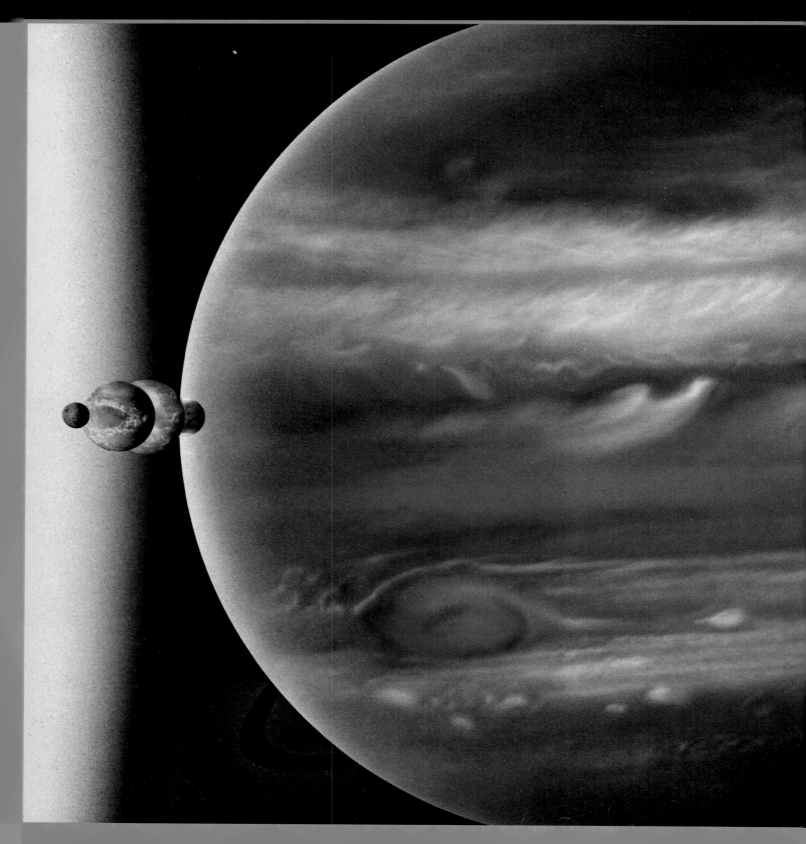

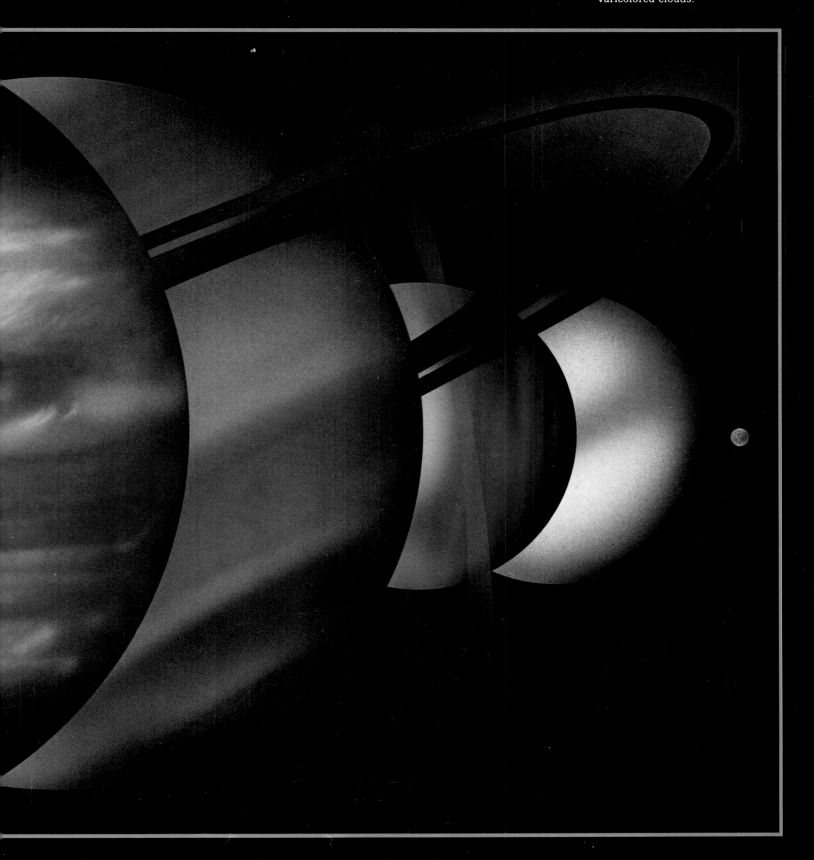

In a composite portrait of the Solar System, the five outer planets, from huge Jupiter *(center)* to tiny Pluto, present an alien architecture of wafer-thin rings, tilted orbits, and varicolored clouds.

ach and every second in the heart of the Sun, hundreds of millions of tons of hydrogen are fused into helium by thermonuclear processes, giving birth to the energy that bathes the Solar System. Rising to the surface of the great orb, this energy streams off into space in the form of light and other radiation, traveling 186,000 miles per second—about 670 million miles per hour. Out past charred Mercury it races, then cloud-covered Venus, and after eight and a quarter minutes and a journey of some 93 million miles, the blue sphere of Earth. Another twenty minutes takes it past the orbit of Mars and a 200-million-mile-wide belt of asteroids, a kind of desolate cosmic borderland. Beyond lie the far planets of the Sun's family, five in all.

With one exception, the outer reaches of the Solar System are the territory of planetary giants—worlds that are primitive, primordial, somehow unfinished. Most possess crushing gravities, powerful magnetic fields, and moons so individual and idiosyncratic as to be worlds unto themselves. Jupiter is first in this realm, both in proximity to the Sun and in size. Orbiting more than 150 million miles beyond the asteroid belt, in a region where the shrunken Sun shines with only one twenty-fifth of its earthly warmth, the immense planet approaches stardom itself. If Jupiter had been endowed with eighty-five times more mass, it could have generated pressure enough to ignite a thermonuclear furnace at its core and blaze up in stellar splendor.

Almost twice as far as Jupiter from the Sun lies the orbit of Saturn, a lightweight colossus that would float in water. Like its mythological namesake, Saturn is an impressive parent, accompanied by seventeen moons, each of indelibly different character. No other planet has so many satellites; even giant Jupiter has one fewer. But it is the luminous multitude of rings—so finely sculpted that they are proportionately thinner than a sheet of onionskin—that most marks the planet.

Uranus, about half the size of Saturn and orbiting a billion miles beyond the ringworld, is distinctive in another way: The planet orbits on its side, its northern and southern hemispheres alternately pointing toward the faint Sun for decades at a time. Aurorae glimmer above the dark pole, wrapped in its forty-two-year night.

Neptune, a seeming twin of Uranus, orbits a billion miles farther on. Apparently girdled by broken rings, this planet and its moons may, like sideways Uranus, commemorate some long-ago violence. One of its moons travels in a direction opposite that of most other satellites, and another loops eccentrically above and below the more normal orbital plane. Both circumstances suggest the ancient passage of a massive object, whose gravity might have wrenched the Neptunian system into its present form—and may also have stripped Pluto away. This stunted afterthought, barely more than a moon itself, travels an orbit that is both the most inclined to the plane of the Solar System and the least circular of all the planets' tracks. Only a feeble wash of sunlight, nine hours old, reaches its rocky surface.

CELESTIAL WANDERERS

To look up from Earth at night has always been to notice the spectacle of Jupiter and Saturn. While the more distant planets are comparative newcomers to human awareness, the wanderings of these two bright points of light once evoked powerful deities and immutable destinies. For the Nile-conscious Egyptians, the moving lights suggested the running lanterns of celestial barges, which the godlike planets sailed along canals flowing from a great river of stars, the Milky Way. In the sixth pre-Christian century, the Greeks, fascinated by geometry, accounted for heavenly events in a different way. They envisioned the universe as a concentric set of fixed, perfect circles. An unmoving Earth resided at the center, and the Sun, Moon, and planets spun on crystalline spheres that were in turn contained within another sphere that held the changeless stars.

But the planets refused to cooperate with this pristine scheme. Observed closely, they seemed now and then to slow, stop, reverse, and then resume their steady course. The planetary loop-the-loop—known as retrograde motion *(pages 16-17)*—was highly disconcerting, impossible to duplicate in a system of concentric spheres. In the end, the strenuous effort to fit observations to theory drove astronomers of the Hellenic world to explain the heavens as a maze of overlapping wheels, circles, and spheres. The arrange-

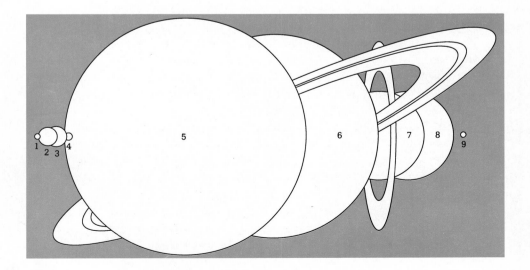

The Sun's nine known planets, shown to scale here and on pages 12-13, divide into two distinct regions. An inner quartet of small rocky worlds—Mercury *(1)*, Venus *(2)*, Earth *(3)*, and Mars *(4)*—clusters near the Sun. Beginning with the giants Jupiter *(5)* and Saturn *(6)*, an outer series includes tipped-over Uranus *(7)* and far-flung Neptune *(8)* and Pluto *(9)*.

ment propounded by Claudius Ptolemy in the second century AD proved to be the most enduring.

Ptolemy's solution was to add to the wheel carrying the Sun and planets a series of smaller wheels, or epicycles, that permitted planets to retrogress without affecting the steady turning of the larger circle. In his celestial Ferris wheel, the cars revolved freely about their hubs, so that a passenger—the planet on its epicycle—would appear to move backward, hover, and then progress with the larger turning wheel. By adjusting the size of the wheels, the speed of rotation, and the radius of the epicycles, the cosmic carnival ride—eventually numbering forty wheels—could produce a great range of motions, all made from the pure stuff of circles. Even this elaborate machine could not account for every observed motion of the planets, but Ptolemy's vision remained sovereign in science for more than a millennium.

COSMIC ARCHITECTS

By the late fifteenth century, in a world being transformed by invention and exploration, the creaky Ptolemaic wheels were primed for dismantling. Amid strong pressure to revive the dormant science of astronomy, a Polish canon in the Roman Catholic church named Mikolaj Kopernik took on the task of translating the ill-defined yearnings of a new age into a universe that worked. Sometime before 1514, Nicolaus Copernicus, as he was called in Latin, circulated a manuscript among a small group of fellow scholars. In it he outlined seven axioms that placed the Sun at the center of the planetary system and the universe and explained that the apparent motions of celestial objects were produced by the Earth's rotation and its revolution about the Sun. He also noted that the fixed stars lay at some great distance from Earth. Though still a many-wheeled machine of large circles and epicycles, the system he devised constituted a fundamental shift in the human perspective. But the cautious canon, perhaps irrationally afraid of general ridicule, resisted efforts by friends and church superiors to have the work published. For more than thirty years, Copernicus held his silence. Finally, on May 24, 1543, a few hours before his death, a printed copy of *On the Revolution of the Heavenly Spheres* was placed beside him on his bed.

Although the densely reasoned tome went virtually unread at the time, it became a philosophical time bomb that would finally explode in the Copernican Revolution a hundred years later. The unwitting standard-bearer of

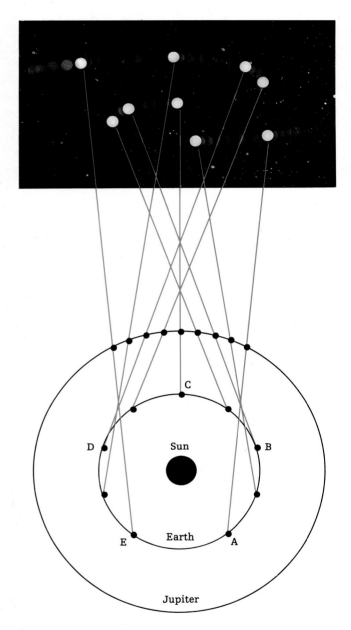

In an illusory maneuver that baffled the early astronomers, Jupiter seems to reverse course from time to time. As shown by the sequential diagram of the paths of Jupiter and Earth at left, the giant planet's lazy pace is what causes the apparent retreat, just as a slow car appears to drift backward from the perspective of a faster one.

When the Earth is at position B, Jupiter lies ahead of Earth and is seen against the stars toward which both planets travel; by position C the two worlds are abreast, and Jupiter is grouped with stars directly opposite the Sun in Earth's sky; by D, the outer world lags well behind, shifting to yet another starry backdrop. Because it travels at a slight angle to Earth's orbital plane, Jupiter traces an S curve, never crossing its previous path.

that revolution was a Dane born three years after Copernicus's death. Tycho Brahe had a well-deserved reputation for flamboyance. For instance, when a duel in his student days cost him part of his nose, he replaced the missing slice with one of gold and silver alloy. But he was also a meticulous observer. As an astronomer, he devoted himself to the measurement of the positions of stars and planets, developing and building instruments of unprecedented size, precision, and expense.

Tycho created an astronomical stir while still in his twenties. The occasion was the appearance of a tremendously bright light in the constellation Cassiopeia in November 1572. Tycho proved that it did not move, like a planet, but was stationary, like a star. (It was, in fact, a supernova marking an explosive stellar death.) Since change was deemed impossible on the celestial sphere that held stars, his proof caused general consternation, but an admiring Danish king gave him the island of Hveen, along with its peasantry, and the wherewithal to build an observatory.

Settling at Hveen, Tycho went on to add still another wild card to the cosmic deck by showing that a great comet seen in 1577 lay beyond the Moon. A comet streaking through that region would thus have to crash through the crystalline spheres carrying the planets. Finally, he developed an alternative to the still-controversial Copernican hypothesis. In Tycho's universe, the planets circled the Sun, and the Sun in turn circled the Earth.

In 1597, his volatile personality putting him at odds with both his king and his tenants, Tycho and his family left their native Denmark. They traveled almost aimlessly in Germany, finally settling two years later near Prague, where Tycho was named Imperial Mathematician to the Holy Roman Emperor Rudolph II. A few years later he was joined in his work by a young man named Johannes Kepler, from the Austrian city of Gratz. Tycho had spent a lifetime carefully gathering the best celestial data of his day. Now he hoped his assistant would use that legacy of precise observation to redefine the universe along Tychonic lines. Kepler would indeed recast the cosmos, but in the mode of Copernicus rather than his mentor.

KEPLER'S LAWS

By the time Kepler arrived in Prague, Tycho had only eighteen months to live. He died after the sudden onset of a malady described only as internal fever, and Kepler took his place as Imperial Mathematician. Kepler spent eight years reviewing the late astronomer's data and searching for order in the apparent contradictions of planets that moved through a perfectly uniform universe at varying speeds and distances from the Sun. In 1609 he published his *New Astronomy,* in which a careful reader could find two momentous laws: first, that planets follow elliptical, not circular, orbits, with the Sun at one focus of the ellipse, and second, that while a planet may move at different velocities at different points along its orbit, its passage always sweeps across the same area within its orbit in the same amount of time. The first law removed the unworkable circle from calculations of planetary orbits. The second accounted for the observed variations in a planet's speed, which, until Kepler, had defied theoretical explanation.

The following year, the universe governed by those laws was illuminated further by the observations of another astronomical titan, working about 500 miles south of Kepler's Prague. In Italy, Galileo Galilei, a professor of mathematics at the University of Padua, had turned his version of a new Dutch invention, the telescope, toward the heavens and set the fuse for the Copernican time bomb. Galileo's *Messenger to the Stars,* published in 1610, declared that the Moon's surface was not—as dogma would have it—perfect, incorruptible, and heavenly but "full of irregularities, uneven, full of hollows and protuberances." The treatise also demonstrated that Venus orbited the Sun. It spoke of fixed stars numerous beyond belief. And it announced a sensational discovery: four "planets"—actually moons—near Jupiter. Proof that Venus, at least, orbited the Sun, and the detection of moons circling Jupiter in the same way one moon circles Earth, made the logic of a heliocentric solar system irresistible. At last, observation overwhelmed belief. Given Galileo's findings, the Ptolemaic universe could not work, nor could the universe be perfect and eternally without change. The available alternatives were the universe of Tycho and that of Copernicus.

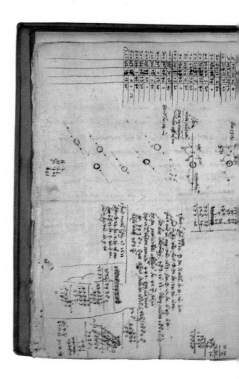

For academicians who had built their careers on the now-discredited Ptolemaic system, these new ideas were dangerous. But Kepler, upon hearing this dissenting chorus, threw his considerable scientific weight to Galileo's side, helping to transform an impossible universe into a Sun-centered one. The work would earn Galileo a summons from the Inquisition in 1632. Kepler, for the most part, escaped severe censure, although his life coincidentally took a turn for the worse. In quick succession, his benefactor, Rudolph II; his wife; and one of his three children died.

While retaining the title of Imperial Mathematician, Kepler was forced to give up the court in Prague for humbler mathematical labors in the provincial city of Linz. Still, he managed to keep developing his bold astronomical perspective. Around 1620 he published a book titled *Harmony of the World,* which contained his third law. In mathematical terms, Kepler declared that the square of a planet's period of revolution varied with the cube of its mean distance from the Sun. In essence, the farther away a planet was from the Sun, or a moon from its planet, the slower its speed along the orbital track. Kepler's second law had explained the variations in a planet's speed along its orbit; this one described the movement of planets in relation to their star. For Kepler, the third law supplied a missing detail. For Isaac Newton, it held the seeds of his universal law of gravitation.

By 1687, roughly half a century after the deaths of both Galileo and Kepler, Newton was able to assemble the structures left by these other architects to show that gravity ruled the universe. Gravity was a force acting at a distance and diminishing with the square of that distance. The orbits produced by this force would be Keplerian ellipses. And the resulting celestial behavior was demonstrated by the Moon's motion around the Earth and the planets' orbits

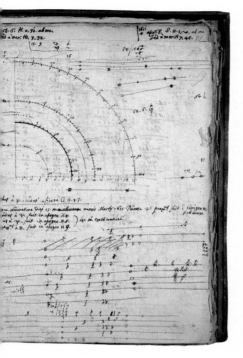

In notes on observations for the night of March 15, 1611, Galileo Galilei recorded a rare alignment of Jupiter and its four larger moons—bodies he called the Medici planets, for his patron Cosimo II de' Medici. Already convinced that the four worlds circled Jupiter, Galileo used the unusual lineup to time the moons' orbits, tracking their motions in the sketches and figures shown here. He reasoned that since all four of them and the planet were poised along a single line of sight, the interval each moon took to return to that position would yield half its orbital period; to allow for observational errors, he counted several revolutions per moon before obtaining average orbit times. A pie-wedge diagram *(above, right)* displays his results; the diagram's lower left corner represents Jupiter, and four arcs, ticked at hourly intervals, show distances and orbit times for the moons now known as Io, Europa, Ganymede, and Callisto. The times Galileo derived fall within half a percent of accepted modern figures.

around the Sun. The dynamics of the universe finally lay within human grasp.

Newton left a technological legacy as well. In 1672, when he had been elected to membership in the Royal Society, England's academy of sciences, Newton had displayed one of his new inventions: a compact reflector telescope. Designed to overcome the shortcomings of lensed telescopes, it would be the vehicle of exploration for the astronomers of the century to come. Such instruments would open the dim regions beyond Saturn to human view.

NEW EYES ON THE SKY

"Seeing," the astronomer William Herschel once wrote to his close friend William Watson, "is in some respect an art, which must be learnt. To make a person see with such a power is nearly the same as if I were to make him play one of Handel's fugues upon the organ." The power he referred to was that of his telescopes, which, while still rudimentary by modern standards, were the world's best in the late eighteenth century.

Herschel himself learned to "see," as well as play Handel, as he seems to have learned everything: by doing it. The musician son of a regimental bandmaster in the Prussian city of Hannover, he grew up debating the ideas of Newton and other scientists, along with music criticism, with his father and brother. According to his younger sister Caroline, their arguments often kept her awake far into the night. In 1753, fifteen-year-old William joined his father and brother in the Hannoverian Guards band.

But Herschel's military career was short-lived. In 1757 the French occupied Hannover during the course of the Seven Years' War, and Herschel and his brother fled to England. Following almost a decade of musical jobs in various English provincial towns, William was appointed organist at the Octagon Chapel in the fashionable spa of Bath, near the bustling port of Bristol. And here his latent fascination with astronomy began to flourish.

Professional interest led him to study the mathematics of music theory. A harmonics text took him to a book on optics, which included a section on astronomical science. When Herschel went to London in 1772 to meet Caroline, who had come to keep house for him, he gave her a tour not of the great city's principal attractions but, she wrote later, of the "optician shops." Herschel was soon making small telescopes from purchased lenses; within half a year he was grinding his own. Despite a heavy schedule of tutoring music, practicing, and composing, Herschel threw himself into the building of the instruments, not even waiting to change his ruffled shirt before pouring pitch for polishing lenses. The hobby inexorably took over the household.

At first Herschel built refractor telescopes, in which light passing through an objective lens is focused at a point, where the image can be examined through a magnifying eyepiece. But, as Newton had pointed out more than a century earlier, the performance of such instruments is limited by the properties of glass lenses, and Herschel soon moved on to the more compact reflector telescopes. In these, light strikes a concave mirror that reflects it to a focal point, where the image can be studied through an eyepiece.

By the time the Herschel household moved to new lodgings in Bath in 1774, William had completed two meticulously polished reflectors that allowed him to obtain images of unprecedented sharpness. Where the era's professional astronomers saw stars as bright smears that when magnified by an eyepiece only became larger smears, he saw points of light, many of them close pairs of stars. Not knowing that most stars are found in pairs, Herschel explained the observations as chance alignments of near stars with more distant ones. Reasoning (wrongly, as it turned out) that the brighter star was closer to the Earth and the fainter one farther away, Herschel hoped that slight variations in the separations of these pairs would show parallax, the apparent displacement of celestial objects by the Earth's changing orbital position. Parallax had been used successfully to calculate the distances to the planets but, because of the very small variations involved, had not worked for stars. With pains, however, any displacement should be more obvious, and he had great faith in his instrument, a 6.2-inch reflector. Only seven feet long, the telescope was easy to set up in the road outside his house in Bath.

One night in late December 1779, as Herschel took a break from his stellar quest to observe the Moon for a change, a passerby asked for a look through the telescope. Impressed by the view, he returned the following day and introduced himself as Dr. William Watson, a fellow of the Royal Society who would become Herschel's lifelong friend and staunch supporter. Watson promptly enrolled Herschel into the then-forming Philosophical Society of Bath, an informal fraternity where dilettantes with a taste for knowledge could mix with the local contingent of scientists. Here Herschel eventually presented some thirty-odd papers on a variety of topics. The last was titled "An Account of a Comet" and referred to something very peculiar Herschel had seen one evening in March 1781.

A CURIOUS COMET

Between ten and eleven o'clock on the night of Tuesday, March 13, Herschel had set up the 6.2-inch reflector, extending his search for double stars toward the constellation Gemini. He noted a "curious either nebulous Star or perhaps a Comet" near the stars marking the left foot of Castor, one of the Twins. Observing the same object four nights later, he took it for a comet, "for it has changed its place." Another reason the object could not be a star was that it appeared to the telescope as a clear disk—like a planet. Herschel then began a series of observations, using a calibrated screw called a micrometer, to measure the strange object's exact position in the eyepiece from night to night. When he passed word of the sighting to Nevil Maskelyne, the Astronomer Royal, and to Thomas Hornsby, professor of astronomy at Oxford, he soon had signals of just how good his telescope was. Hornsby could not find the object through his instrument, and Maskelyne could not detect its movement across the sky.

Still, there was something very peculiar about the so-called comet. Maskelyne wrote in April to thank Herschel for communicating information on

THE PLANETARY PANTHEON

From the beginning, the far planets and their moons were identified with mythic Roman and Greek deities. Saturn, one of the sons of Uranus, overthrew his father to rule his race of immortals—only to be subdued in turn by the thunderbolts of his own son, Jupiter.

The sixteen Jovian moons are mostly named after objects of Jupiter's amorous campaigns, including Io, Europa, Callisto, Leda, and Ganymede, a boy from Greek mythology. But there are exceptions, such as Metis, a goddess who caused the child-devouring Saturn to spit out Jupiter's siblings, and Amalthea, a goat whose milk nourished the infant Jove.

Saturn's largest satellite, Titan, was named for an early race of immortals descended from Uranus. Its sixteen companions include the Titans Hyperion and Iapetus. Dione, Phoebe, and Rhea were Titan women; Tethys, a sea goddess, was Dione's mother.

The five Uranian moons discovered from Earth since 1787 were cast as characters from English literature— Miranda, Ariel, Oberon, and Titania from Shakespeare, Umbriel from Alexander Pope. These authors also provide the names given to the ten smaller satellites seen by *Voyager 2.*

Neptune was given the name of Jupiter's brother, the powerful god of the sea, and the larger of its two known moons became Triton, Neptune's fish-bodied son. The other was named Nereid for the sea nymphs lured by Triton's conch-shell music.

The discovery of a ninth planet triggered a torrent of mythical nominees for its appellation, with the decision going to Pluto, another of Saturn's sons. As lord of the underworld, Pluto travels in the close company of the grim ferryman of the dead, Charon, the name given the planet's twinlike moon.

"the present Comet, or planet, I don't know which to call it." Later that same month, Charles Messier, a French astronomer noted for his comet hunting, wrote Herschel that "this comet" bore little resemblance to any of the eighteen Messier had observed. Moreover, all efforts to shape the object's observed movement into the kind of nearly parabolic orbit followed by most comets failed miserably. The calculated orbit would coincide with the actual path of the object for only a few days before diverging.

So indelible was the age-old belief in a solar system of six planets that the possibility that this new object might be a seventh did not immediately occur to Herschel, or to many others. Finally, a Russian astronomer named Anders Lexell, who was visiting in London at the time, concluded from the object's clear, disk-like shape and its slow, steady movement that it had to be a planet, an idea that won general acceptance after he announced his calculations of a probable orbit.

If Herschel's discovery was quickly accepted by the astronomical community, his methods were not. When he presented his findings to the Royal Society on April 26, he reported matter-of-factly that he had used a telescope whose resolution exceeded the prevailing state of the art, at magnifications that were unthinkable. His critics were inclined to treat him as a pretentious amateur who made irresponsible claims for his homemade instruments. Their skepticism was reinforced by the fact that Herschel had miscalculated the distance to the object, which proved to be farther away than his micrometer readings showed. (Although Herschel's micrometers were faulty, a 1924 examination of his extant instruments found that he had, if anything, underestimated the magnifying power of his eyepiece.)

His friends, on the other hand, helped Herschel put the new planet to work on his behalf. Herschel's name was mentioned to the king, along with the suggestion that the newly discovered planet be named George III. Commanded to appear at Windsor Castle in the spring of 1782, Herschel introduced the king to telescope astronomy, and later asked for royal support. George grant-

GRAVITY AND MASS

When Isaac Newton pondered the properties of gravity in the seventeenth century, he established laws that still serve planetary scientists today. In the first part of his theory of universal gravitation, Newton stated that the force of attraction between two bodies is directly proportional to their masses. (Mass is a measure of the amount of matter in a body.) Any two objects will attract each other, and the strength of that attraction at a given distance depends upon multiplying the two masses.

In the second part of his theory, known as the inverse-square law, Newton said that the force of attraction diminishes inversely according to the square of the distance between two objects. For instance, when a spaceship leaving a planet doubles its distance from the planet's center, the pull of gravity on the ship is reduced to a quarter of its original strength; when the ship increases its distance three times, the attraction drops to one-ninth, and so on.

A planet's gravitational field can be neatly compared to a sloping well, with the planet at the bottom *(right)*. The depth of the well determines the energy required to escape the field. Because a very massive planet such as Jupiter would possess a deep well with steep sides, a spaceship leaving that planet would work its way up with great effort at first, then travel with increasing ease as it neared the edge of the well and emerged onto a nearly horizontal plane. The same ship, climbing out of smaller Neptune's well, would only travel one-sixth the distance and reach the plane with a relatively small expenditure of energy.

When it comes to falling or orbiting bodies in space, however, distance, not mass, determines an object's gravitational acceleration. Gravity attracts each particle of mass individually; thus, a ten-pound weight and a one-pound weight falling from the same distance—provided a force such as friction did not slow the larger body—would fall as though they were eleven separate one-pound objects and would hit the ground simultaneously. Placed in the same orbit, both masses would move at the same velocity. Thus, if Jupiter were placed in Earth's orbit, it would circle the Sun at the same speed as the less-massive planet.

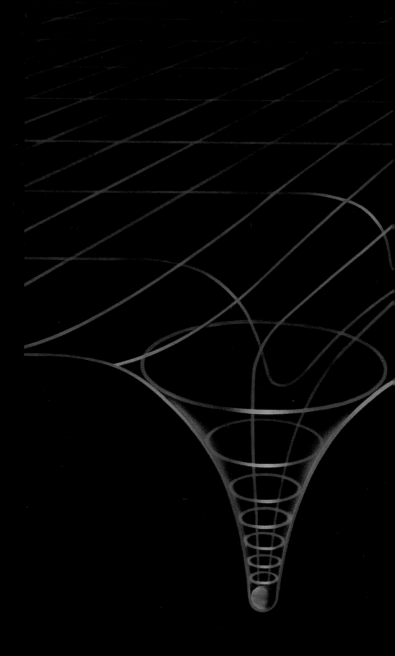

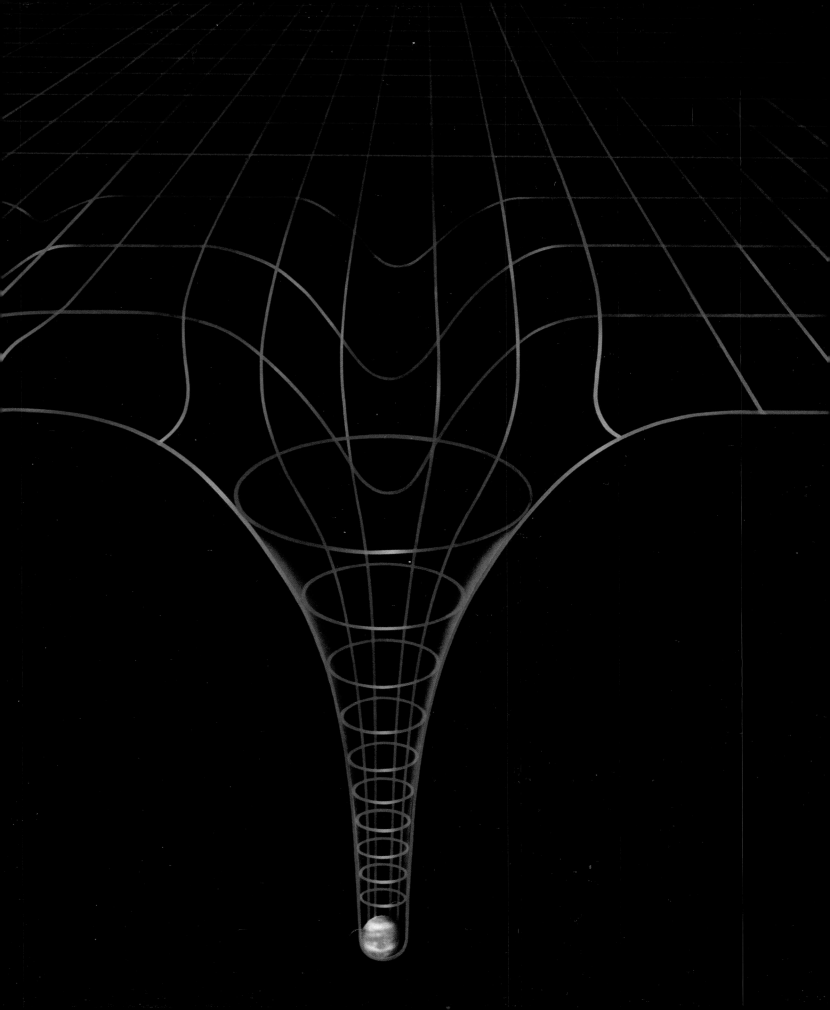

Gravity and Motion

Newton built on the conclusions of German astronomer Johannes Kepler to explain why planets travel in elliptical orbits, moving faster when closer to the Sun. Knowing that objects in motion continue in a straight line at a constant velocity unless acted upon by a force—that is, they have inertia—Newton decided that gravity must be acting upon the planets to keep them circling the Sun.

The way motion and gravity interact to shape orbits can be shown using a model of a gravitational well *(right),* with an attracting mass hidden at the bottom. Traveling along the well's sides are three bodies. Each has kinetic energy due to its own motion; each also has a certain amount of potential, or stored, energy from the gravitational force pulling on it. Although each body's total energy remains constant, when it moves down the well some of its potential energy changes to kinetic energy, causing it to move faster (shown by overlapping balls) as it nears the attracting mass. Similarly, bodies in closer orbits *(gold)* move faster than those in more distant orbits *(purple).*

The balance between kinetic and potential energies determines the path a body follows. When a body's potential energy exceeds its kinetic energy, it assumes an elliptical or near-circular orbit around the more massive object, as planets and moons do. However, if its potential energy is much larger, the body spirals into ever-tighter orbits, eventually crashing into the attracting mass. And, if the kinetic energy of a body exceeds its potential energy, the body will escape in a hyperbolic curve *(blue),* never to return.

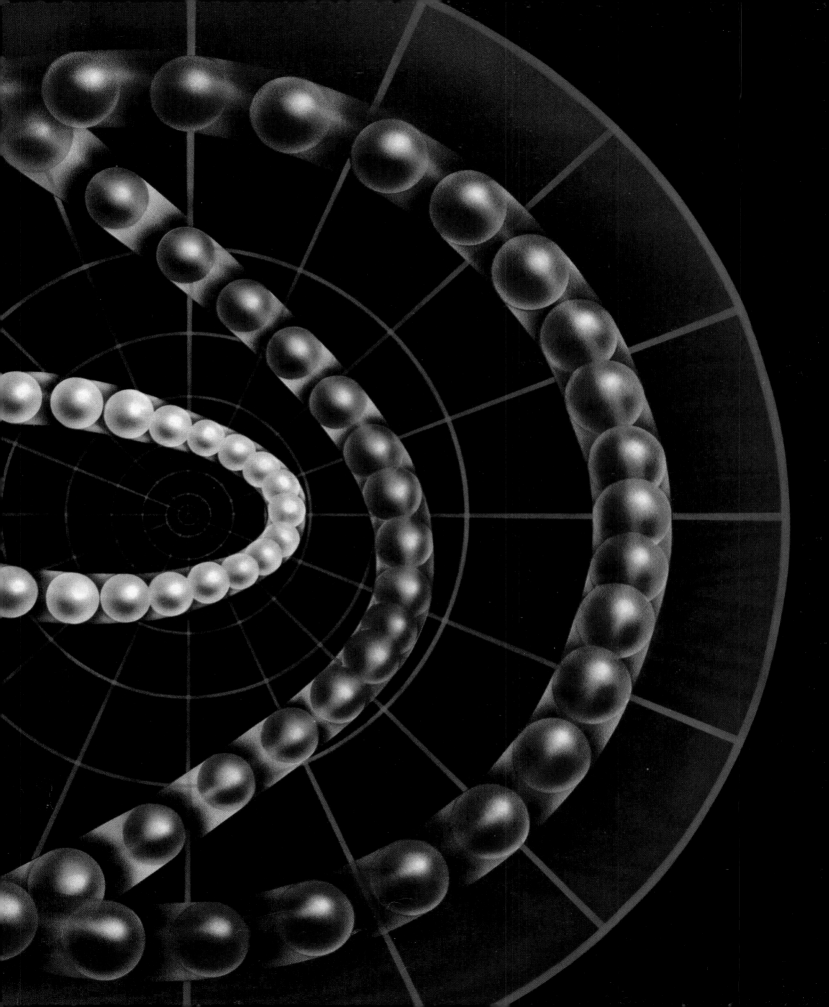

ed him an annual stipend of £200 (plus a £50 stipend for Caroline) and named Herschel "Astronomer under Royal Patronage," whose sole duties were to let the royal family look through his telescope. Although the wages were barely half what his music earned in Bath, Herschel quickly accepted. He moved to a house within two miles of Windsor Castle, near London, pleased to be able to compare instruments with the professionals at court but not much taken with high society. "I would rather be polishing a speculum," he wrote Caroline, referring to the reflecting mirror of his telescopes. To make ends meet, Herschel built and sold telescopes. Then, in 1788, he married the wealthy widow of a London merchant. The amateur hobbyist had become a rich and famous astronomer.

Meanwhile, the name of the new planet remained unresolved. Herschel admirers in France promoted the appellation Planet Herschel; others pushed for variants of Neptune or Uranus. Herschel himself continued to call it the Georgian Planet, and the general view in his adoptive land was to see the find in terms of British conquest. Said one scholar, "It is true we had lost the terra firma of the Thirteen Colonies in America, but we ought to be satisfied with having gained in return by the generalship of Dr. Herschel a terra incognita of much greater extent."

For the next sixty years, the discovery bore three different names. But in the end, the tradition of mythology won out, and the planet became Uranus.

BLOCKING STARLIGHT

Even before the space age brought close-up images of the outer planets, earthbound scientists gleaned nuggets of information about these distant worlds by making use of an event known as occultation (from the Latin *occultare*, meaning "to conceal"). As shown here and on the following pages, the phenomenon occurs when one celestial object temporarily blocks all or part of another from view—as, for example, when the Moon occults the Sun during a solar eclipse.

When a planet occults a distant star *(right)*, scientists can learn whether the planet has an atmosphere by observing whether the star vanishes abruptly or dims gradually. From the changes in the rate at which the star's light fades and the degree to which it is bent, or refracted, observers can calculate the atmosphere's depth and make educated guesses about its chemical composition.

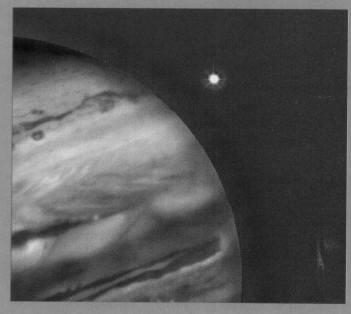

As a planet—Jupiter, in this case—travels along its orbital path, it may pass in front of, or occult, a distant star.

NEPTUNE'S NET

Hardly had Uranus been spotted and its orbit calculated than some astronomers began to puzzle over its behavior. The longer they watched Uranus on its eighty-four-year journey around the Sun, the less its path followed the one predicted by Newton's universal law of gravitation. Perhaps, scientists speculated, the gravitational rules became less rigid at great distances. Or perhaps there was yet another planet, trailing its gravitational net into the Uranian orbit. The riddle had a gravitation all its own and by the 1840s had attracted two gifted, but very different, young mathematicians: John Couch Adams in England and Urbain Jean Joseph Leverrier in France.

Born in 1819, Adams was the son of a tenant farmer in the East Cornwall town of Launceston. He made his talent for mathematics obvious early on, teaching himself calculus, number theory, and "mechanics," or physics. Upon entering Cambridge University, where he studied mathematics, he set his own high standards, reprimanding himself for indulging overmuch his fondness for astronomy. One day in June 1841, while browsing in a Cambridge bookstore, Adams came across a report describing the current quandary over the peculiar motion of Uranus. The paper had been written by George (later Sir George) Airy in 1832 while he was still a Cambridge professor, before beginning nearly half a century as Astronomer Royal. Intrigued, Adams decided to study the peculiarities of Uranus's orbit to determine whether they could be

The star's light begins to dim as Jupiter draws near, suggesting the presence of a gaseous envelope around the planet.

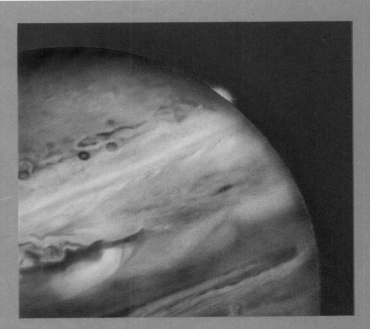

The length of time it takes for the star to vanish behind the planet gives clues to atmospheric depth and composition.

attributed to an unseen planet beyond it—and if so, to discover that planet by plotting its orbit in turn.

Armed with data on Uranus's position supplied by James Challis, professor of astronomy and director of the Cambridge Observatory, the twenty-two-year-old Adams began spending his holidays working on the requisite long calculations. By October 1843 he had a preliminary solution for where the presumptive eighth planet must be, and two years later he had refined his calculations further. By this time, having performed brilliantly in mathematics at Cambridge, Adams was a fellow at the university's St. John's College. He showed his computations to Challis, who in turn urged him to present the work to Airy. A few weeks later, en route to visit his parents in Cornwall, Adams stopped by Airy's Greenwich Observatory. But the Astronomer Royal was out of the country then and was unavailable again when the young man tried to see him on his way back to Cambridge. Adams left the calculations for Airy to review, but he heard nothing for some time. Finally, in November 1845, a note came from Airy. With little in the way of preamble or encouragement, the Astronomer Royal asked merely whether Adams's solution explained discrepancies in both the distance and motion of Uranus. Feeling snubbed by Airy's earlier neglect, Adams did not reply.

As it happened, Airy was skeptical that the irregular behavior of Uranus could be explained any time soon—if at all. His cool reception to a theoretical

MEASURING A MOON

Although it is rather rare for the outer planets to occult bright stars, they often block the view from Earth of their own orbiting satellites *(right)*. Scientists make use of this event to gauge a moon's diameter.

As the moon begins to pass behind its planet, it reflects progressively less and less light—registered by light-measuring devices, called photometers, attached to a telescope—until it vanishes completely. When astronomers have established a moon's orbital speed through other methods, they can calculate its diameter from the elapsed time between its first dimming as it enters the shaded area, known as the penumbra, and its final disappearance: The larger the moon, the longer this disappearing act takes.

In this overhead view, a moon of known orbital speed approaches the penumbra, a shadowy area on either side of a planet.

analysis from a young unknown was perhaps to be expected. A greater surprise was the indifference shown by James Challis, the original source of Adams's data and someone who must have been familiar with Adams's reputation at Cambridge. Inexplicably, Challis did not even bother to train the university's twelve-inch Northumberland telescope on the patch of sky where Adams predicted the new planet could be found.

THE FRENCH CONNECTION

Across the English Channel, unaware of Adams's work, Leverrier was also in hot pursuit of the causes underlying Uranus's anomalous orbit. Born in the town of Saint-Lô in 1811, Leverrier was the son of a local government official who wanted the best education for his child. The boy flunked the notoriously tough entrance exams to the École Polytechnique in Paris but later, after attending another university, more than justified his father's faith in him by winning high honors in a prestigious exam administered by the École Polytechnique in 1831.

After beginning his career as a chemist, Leverrier became a lecturer in astronomy at his alma mater in 1836 and an assistant astronomer at the Paris Observatory, where in 1854 he would succeed Dominique Arago as director. Astronomy drew upon Leverrier's strengths: ambition and a willingness to tackle any calculations, no matter how demanding. By the time the puzzle of

The leading edge of the moon enters the shadow, and the timing of its eclipse begins.

When the last of the moon disappears into the penumbra, timing stops and the moon's diameter may be calculated.

Uranian behavior was brought to his attention by Arago, Leverrier had gained recognition as a consummate mathematical analyst.

In June 1846, eight months after Adams's prediction of the missing planet's orbital position, Leverrier published a paper that eliminated all other explanations for the strange behavior of Uranus. There had to be an undiscovered eighth planet. Moreover, the planet's orbit had to lie a certain distance from Uranus. If it were too close, its gravitation would tangle with Saturn's orbit, which showed no signs of such disturbance; if it lay too far away, it would have to be so massive for its gravitation to affect Uranus that it would also trouble Saturn. Therefore, according to Leverrier, the perturbing planet should be twice the distance of Uranus from the Sun. He calculated a position on the sky that he believed to be accurate within ten degrees of arc—about twenty times the apparent width of a full Moon.

The collective reaction of French astronomers amounted to dignified hurrahs for this new analytical triumph by an eminent countryman. But, inex-

A URANIAN BONUS

In 1977 astronomers studying Uranus as it occulted a bright red giant star learned that Saturn is not the only planet in the Solar System possessed of rings. Several pairs of photometers were set up at Earth, one of each pair tuned to red wavelengths to record the star's light, the other tuned to record blue light from blue-green Uranus. To be sure that the readouts were re- cording real effects at Uranus, they were checked against each other: Matching blips would indicate interference from an intermediate source such as Earth's atmosphere. Just before Uranus hid the star, the red-sensitive readout flickered, then dipped deeply when the star disappeared behind the planet, then flickered again in a mirror image pattern as the star reappeared *(right)*. Since the blue-sensitive readout held relatively constant, the flickering implied a series of rings, each blocking the star's light in turn.

plicably, none tried to confirm it with a telescope. In England, George Airy expressed admiration for Leverrier's tour de force. During late June 1846, Airy wrote to Leverrier and asked him the same questions he had asked Adams about how well the hypothesized planet solved the problems of Uranus's orbit. However, he neglected to mention that Adams had achieved essentially identical results the year before. Leverrier, unlike Adams, replied to Airy's apparent satisfaction.

A few days after sending his letter to Leverrier, Airy attended a scientific meeting at Greenwich Observatory, where he did bring up this close coincidence between the work of Adams and Leverrier. Also in attendance were Challis and Sir John Herschel, the astronomer son of the discoverer of Uranus. Herschel was impressed enough to cause a furor two months later by publishing a letter claiming the discovery for Adams. Challis, on the other hand, seemingly lost interest in the belated efforts of the Astronomer Royal to champion the quest for the new planet. Asked by Airy in July to begin a search,

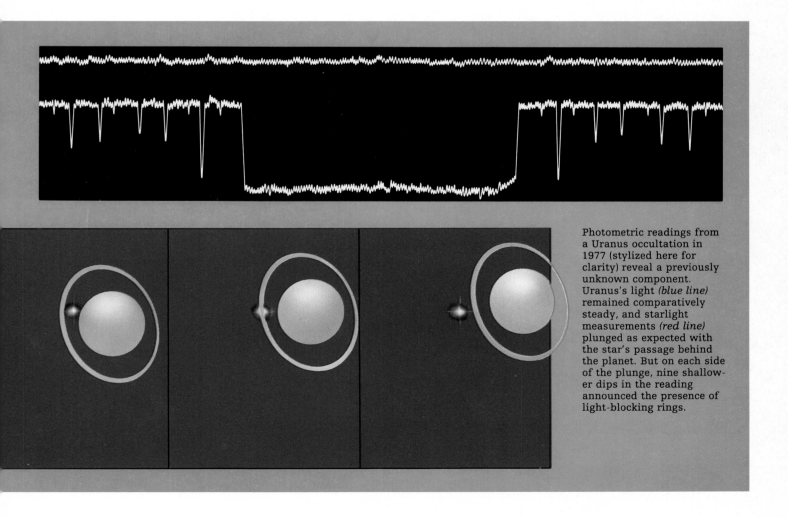

Photometric readings from a Uranus occultation in 1977 (stylized here for clarity) reveal a previously unknown component. Uranus's light *(blue line)* remained comparatively steady, and starlight measurements *(red line)* plunged as expected with the star's passage behind the planet. But on each side of the plunge, nine shallower dips in the reading announced the presence of light-blocking rings.

he proceeded without apparent enthusiasm. At the end of August, another paper from Leverrier arrived, predicting the new planet's size and indicating that it would appear as a disk, not a stellar point of light, to the telescope.

The Frenchman himself, meanwhile, was frantically trying to interest someone, anyone, in aiming a powerful telescope at his predicted position, but to no avail. Then, shifting papers on his cluttered desk, he noticed a doctoral thesis sent to him by an admiring young assistant at the Berlin Observatory. Perceiving this as a last straw to grasp, he wrote Johann Gottfried Galle, begging pardon for his earlier neglect and tossing out desultory praise of the young man's paper before asking him to look for the planet.

Leverrier's letter reached Galle on September 23, 1846. That same night, working with Heinrich d'Arrest, a student assistant, he trained Berlin's nine-inch Fraunhofer reflector on the region of the sky where the eighth planet was predicted to be. As Galle called out the positions of stars from the telescope, the assistant checked them against objects mapped on star charts. Suddenly the student exclaimed, "That star is not on the map!" Leverrier's predicted coordinates were right on the mark.

In England, Challis continued his sluggish pursuit of the new planet, and on September 29, he saw one star that was suspiciously disklike. Instead of following through on the observation, however, he put the telescope aside and went to bed. The next day, September 30, he learned that the observatories in Paris and Berlin had found Neptune. As it happened, Challis had thrown away his chance at immortality much earlier; if he had checked his August observations, he would have come upon two sightings of the new planet.

PLANET X

For Adams and Leverrier, Neptune was the solution. For the generations of astronomers that followed, it became the problem. Once Neptune's path had been established with reasonable accuracy, it was clear that the eighth planet did not explain all the discrepancies between the predicted and actual orbits of Uranus. Trained by the Neptune experience, turn-of-the-century astronomers quickly concluded there must be still another planet, and they set out to discover it. Two Americans led the way.

Percival Lowell, scion of a wealthy Boston family, devoted himself to astronomy after seeing an Italian astronomer's report of *"canali"* on Mars. Using his own money, Lowell built an observatory at an elevation of 7,200 feet in the mountains near Flagstaff, Arizona, and began observing the planets through the clear, dry air in 1894. His twenty-four-inch instrument easily picked out the Martian "canals," along with numerous other characteristics he mistakenly took for signs of life. Lowell's claims for an inhabited Mars were scorned by most astronomers—but not all. William Pickering, whose older brother Edward was director of the Harvard College Observatory, also thought he saw oases on Mars, and even signs of life on the Moon. Pickering had helped Lowell establish his Flagstaff site, and both men became obsessed with the idea of finding a trans-Neptunian planet.

The Earth's atmosphere shields it from most radiation, preserving life but limiting the ground-based study of planets and other bodies mainly to their emissions of light and radio waves. Both are forms of electromagnetic energy, differing only by wavelength. The full electromagnetic spectrum is represented below. Stylized waves indicate the altitudes at which other parts of the spectrum are absorbed; their detection requires high-flying devices often carried by rockets or satellites.

Lowell argued that the search should be based on only the observed deviations in the orbit of Uranus, since Neptune's orbit was not precisely known. As he later wrote in his memoirs, "The sole road to any hope of capture lies through the methodical application of laborious analysis." The remoteness of the hypothesized body, which Lowell dubbed Planet X, also presented a formidable challenge to observers. In fact, the quest he started in 1905 would be longer and more difficult than anyone ever dreamed at the beginning, despite the major advances in instrumentation and technique then occurring in astronomy.

Photographic plates able to gather light for lengthy periods of time were displacing the human eye at the prime focus of telescopes, opening the possibility of completely surveying the heavens. Lowell recognized that no matter how well his calculations predicted the position of the unknown planet,

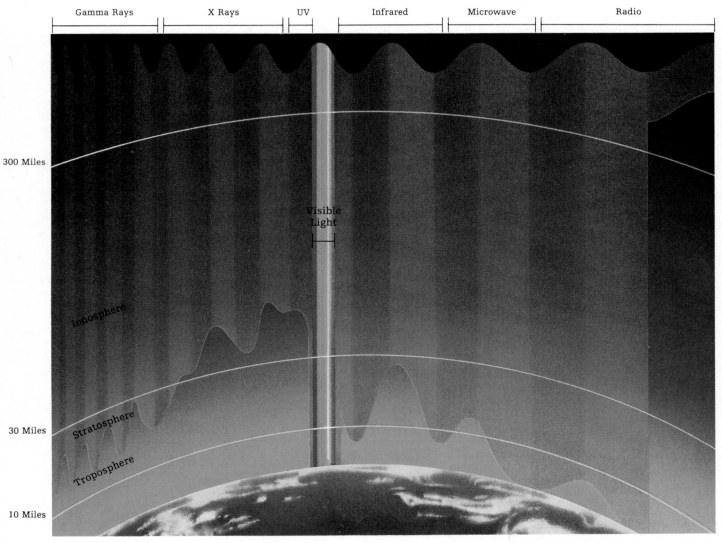

Gamma Rays | X Rays | UV | Infrared | Microwave | Radio

Visible Light

300 Miles

Ionosphere

30 Miles

Stratosphere

Troposphere

10 Miles

photography was his only hope of finding this single, exceedingly faint, moving point of light among thousands of fixed stars. Like frames in an animated film, serial photographs of a given patch of night sky might capture the elusive motion. But after several years of unsuccessful searching, Lowell realized his methods were inadequate to the task. He had been comparing two photographic plates by placing one atop the other and was using a hand lens to look for a mismatched object. In 1911 he ordered what astronomers call a blink comparator for his Arizona observatory. This device allows quick comparisons of two successive photographic plates of the same patch of sky; any object that has changed position between photographs will "blink," or appear to jump from one position to another, while more distant stars remain steady.

For the next five years, the observatory staff systematically photographed the Arizona skies, compared an endless succession of frames, and still came up empty-handed. The sixty-one-year-old Lowell was understandably disheartened by his continued failure to find the hypothetical planet. Late in 1916, he died of a stroke in Flagstaff.

Because the mystery body troubled the orbit of Neptune, Pickering named it Planet O, the letter in the alphabet following N. His approach to finding the planet was a quicker, simpler, empirical one based on the observed positions of Saturn and Neptune as well as Uranus. Drawing graphs of the differences between the actual and predicted positions of those planets, Pickering looked for telltale signs of their being pulled ahead or dragged behind by the gravitational touch of an unknown body. In 1919 Pickering predicted that Planet O lay near the border between the constellations Taurus and Gemini.

The work of both Lowell and Pickering attracted the interest of Milton Humason, an astronomer at California's Mount Wilson Observatory. During late December of the year 1919, Mount Wilson astronomers swung their ten-inch Cooke telescope toward the calculated coordinates on the sky. But they found nothing. Convinced that Planet O did not exist, Pickering gave up the pursuit for nearly a decade.

When the search resumed, in the late 1920s, the Lowell Observatory was again the center of action. Astronomers at the Flagstaff site, now led by a brilliant spectroscopist by the name of Vesto Slipher, had just built a new, thirteen-inch refractor telescope. They believed their device could find what the Cooke telescope might have missed, but they knew that the hunt would be both tedious and demanding. What Slipher needed was an observational astronomer of robust constitution, someone who could endure all-night observing runs, winter and summer, in an opened dome perched more than a mile above sea level. Fortune handed him the ideal candidate in the person of a Kansas farm boy.

By all accounts, Clyde Tombaugh had never wanted to be anything but an astronomer. At the age of twelve, he had looked through an uncle's homemade three-inch telescope and from then on read voraciously in astronomy. At Burdette High School, few classmates understood why the scant time Tombaugh had left after homework and a heavy load of chores should be spent

Magnified portions of the paired photographs that confirmed the existence of Pluto record the tiny world's shift in position (arrows) from January 23, 1930 (top), to January 29. Only its slight motion singles out Pluto from the swarm of background stars, most as bright or brighter than the planet.

alone at the eyepiece of a homemade telescope. They called him "Comet Clyde," writing in their yearbook, "He will discover a new world."

After graduation there was no money in the family till for a college education. Tombaugh spent his days working the Kansas wheat fields, his nights working the dark, star-spattered sky, using a nine-inch lens he had painstakingly ground himself in an underground chamber he had dug near his house. This subterranean workshop offered naturally stable temperatures, essential to prevent thermal changes from deforming the mirror while it was being shaped. When completed, his scope rested on a jury-rigged stand made from a discarded cream separator, the shaft of a 1910 Buick, and various parts salvaged from a straw spreader.

ABANDONING THE PLAINS

In the fall of 1928, Tombaugh sent to Flagstaff sketches he had made of Lowell's beloved canals on Mars and of the banded face of Jupiter. Impressed with the drawings' fidelity, Slipher invited Tombaugh to join the Flagstaff team as an assistant, on a trial basis. The following January, the twenty-two-year-old, who had never before been out of the American midwest, committed his fortunes to unknown territory, arriving in the Arizona city without enough money in his wallet for a return fare.

Homesick, hemmed in by the yellow pines of the unfamiliar, rust-colored mountains, Tombaugh spent his spring nights with the observatory's new thirteen-inch instrument, taking hour-long photographic exposures of Gemini, one of the regions favored by Lowell for finding Planet X. Unknowingly, he was actually recording what he searched for, a tiny object hiding in the darkness beyond Neptune. But Slipher, eager for a quick find, saw nothing on his rapid scan of the plates in the blink comparator. The young apprentice could not help noticing the disappointment in Slipher's face as the older man rose from the scanner. But Tombaugh had also noticed his director's mistakes, and he set about correcting them.

Assigned by his discouraged boss to blink the plates by day as well as exposing them at night, Tombaugh soon realized that for optimum results the photography should be done when the search area lay in opposition—that is, when the Earth was between the area to be photographed and the Sun. At that time, he reasoned, the dim object would be most brightly illuminated, and it would also show the

Clyde Tombaugh peers into the blink comparator he used to discover Pluto in the images at left. The device alternates its operator's view between two fourteen-by-seventeen-inch photographic plates taken days apart; visible here is the left plate, which reflects Tombaugh's hand. As the photographs are compared, stars remain still, but planets, asteroids, and other moving bodies change position to produce a noticeable "blink."

maximum amount of retrograde motion, seeming to slip backward. In fact, Lowell had suggested taking search photos at opposition, a hint most of his successors failed to note.

Tombaugh also came to grips with some operational problems. The fourteen-by-seventeen-inch glass photographic plates were brittle and tricky to handle. Moreover, processing them introduced chemical defects—chance deposits of silver and particles of dirt—that hampered the young man's efforts to find a blinking dot. He decided to scan only photographic negatives to keep processing to a minimum.

A more challenging problem had been the sudden appearance of a legion of blinking objects that turned out to be asteroids scattered among the hundreds of thousands of stars on each photographic plate. This asteroidal nuisance went away when he searched at opposition, where their rapid motion during the long exposures smeared their images with easily detected tails. The tails would distinguish them from true candidates for Planet X, which, being more distant and slower, would show less movement—and no appreciable smearing—between frames.

On February 18, 1930, Tombaugh sat comparing photographs he had made a month earlier of Gemini's star fields, when that constellation lay at opposition. In the late afternoon, with a quarter of the plates scanned, two faint images, one on each of the plates in the comparator, blinked at him. They had not moved much, changing position only an eighth of an inch between the two images, but they had definitely moved. Noting the time on his watch—very near four o'clock—he kept his excitement in check while he carefully verified that the amount and direction of shift, and the faintness of the image, corresponded to what was expected of a trans-Neptune planet. Only then did he step from his office to notify his superiors.

Strict silence about the find was enforced at the observatory, which wanted first crack at calculating the orbit of this dim object, so distant that it could not be resolved into the familiar disk of a planet. And yet, it moved like a planet, traveling precisely as it should across the sky. On March 13, 1930, the anniversary of Herschel's discovery of Uranus and the seventy-fifth anniversary of Percival Lowell's birth, the observatory took the plunge, announcing the discovery to the world. Around the globe, telescopes swung toward this new member of the solar family, a frozen little globe whose orbit extended the Sun's known influence nearly a billion miles beyond Neptune. In May 1930 the ninth planet was named Pluto.

There the matter rests—or poises. As it turns out, Pluto is much too small to produce the perturbations seen in the orbits of Uranus and Neptune. If still another Planet X is out there, it is unlikely that any earthbound eye will see it. Most astronomers today believe that such an undiscovered body would lie too far away for its reflected light or apparent motion to be caught by traditional instruments. But this further mystery may yet be solved—as many of the mysteries of the far planets have been—by a robot explorer, drifting by chance into the gravitational province of a still more distant wanderer.

EXPEDITIONS OF DISCOVERY

Earth's view of the outer Solar System ranges from fair to dismal; the distances are simply too great for astronomers to discern much detail. Thus, for more than three decades, automated spacecraft have been sent ever deeper into space, visiting the planets beyond Mars one after another. By 1990 only Pluto will remain unexplored. Four ships have already made the journey past Jupiter. These difficult voyages of discovery require precise timing and pinpoint accuracy. Mission controllers must launch their probes from the moving platform of Earth at planetary targets moving in their own right. Moreover, to sail the long reaches of interplanetary space, the probes must harness the gravity of the Sun and the planets themselves. Each vehicle is first launched into an eccentric orbit around the Sun that puts it nearly on a collision course with Jupiter *(pages 38-39)*. As the probe draws near, the planet's gravity changes the ship's trajectory, sending it on to its next target.

Pioneer 10, launched in 1972, was hurled by Jupiter out of the Solar System. A year later *Pioneer 11* rounded Jupiter on its way to Saturn. *Voyager 2*, the champion explorer, left Earth in 1977: After its Jovian encounter in 1979, it was tossed to Saturn, whose gravity was used to set the craft on a course for Uranus in 1986 and Neptune three years later. *Voyager 1*, launched a month after *Voyager 2*, overtook its sibling en route to Jupiter and met with Saturn before heading toward the edge of the Solar System.

The newest mission, Galileo, is the first designed to perform a long-term, detailed study of one of the outer planets. Scheduled to head for Jupiter in 1989, the heavy craft will spend six years on a complicated course that provides one gravitational boost from Venus and two from Earth. Moving among the giant planet's moons for two years, *Galileo* will give scientists their best chance yet to plumb the mysteries of one of Earth's distant neighbors.

A Family of Deep-Space Explorers

In the course of three generations of technology, deep-space explorers sent from the Earth have become increasingly sophisticated, but they retain a family resemblance in their largest features. A typical probe is dominated by its large dish-shaped radio antenna, three to five yards in diameter, that links it to its terrestrial base, hundreds of millions of miles away. The antenna is mounted on a central bus—the body of the spacecraft—which contains control electronics, computers, and some scientific instruments. At the end of a long boom projecting from the bus is a small plutonium electrical generator. (The remote Sun is so dim that solar energy cells are not efficient.) Other booms hold scientific devices that must be kept away from the generator's radiation and from the equip-

ment on the bus. Special cameras may be mounted on booms or the bus.

Because complete precision is impossible to achieve in a launch from Earth, all the spacecraft employ small rocket thrusters to rotate the vehicle and to execute periodic course corrections that refine the ship's trajectory. *Galileo,* designed to enter a long-term orbit around Jupiter, also carries a retrorocket to slow the vehicle so it can be captured by the planet's gravity. In addition, *Galileo* is equipped with a separate capsule intended to measure the temperature and examine the composition and structure of Jupiter's atmosphere. Eleven scientific experiments aboard the craft will allow scientists to gather information about Jupiter and its moons in unprecedented detail.

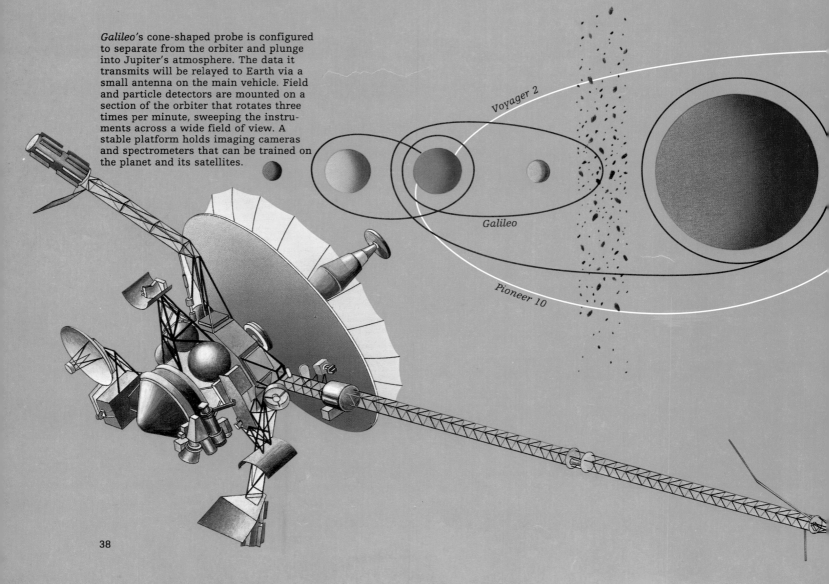

Galileo's cone-shaped probe is configured to separate from the orbiter and plunge into Jupiter's atmosphere. The data it transmits will be relayed to Earth via a small antenna on the main vehicle. Field and particle detectors are mounted on a section of the orbiter that rotates three times per minute, sweeping the instruments across a wide field of view. A stable platform holds imaging cameras and spectrometers that can be trained on the planet and its satellites.

Voyager 2

Galileo

Pioneer 10

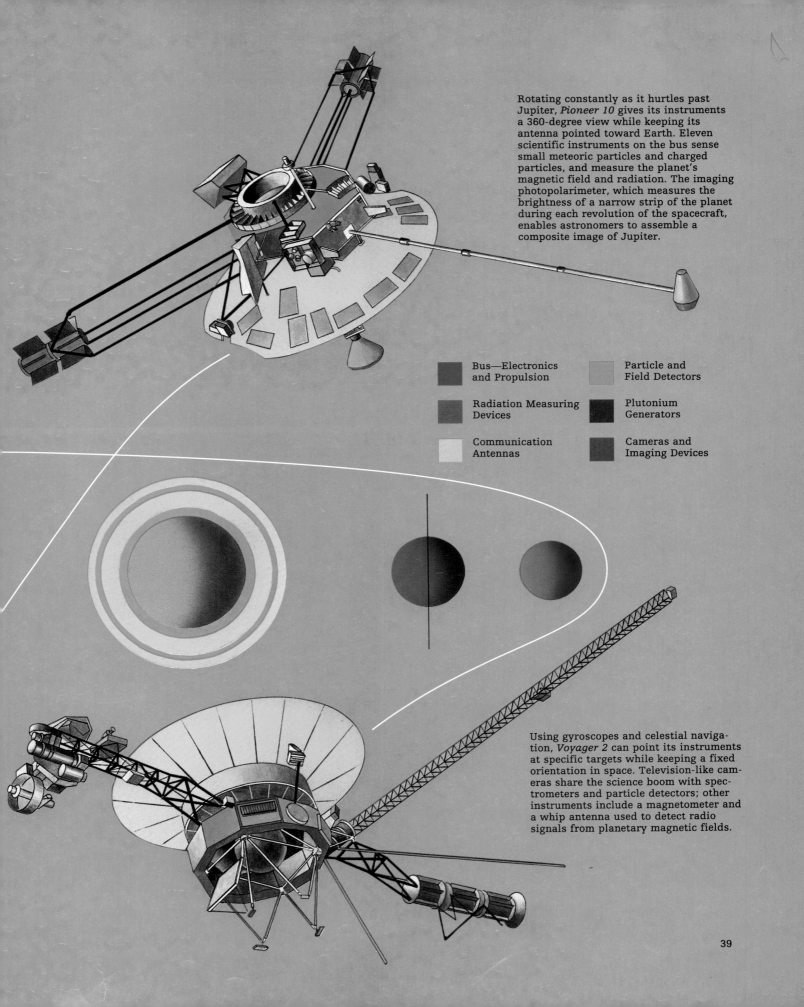

Rotating constantly as it hurtles past Jupiter, *Pioneer 10* gives its instruments a 360-degree view while keeping its antenna pointed toward Earth. Eleven scientific instruments on the bus sense small meteoric particles and charged particles, and measure the planet's magnetic field and radiation. The imaging photopolarimeter, which measures the brightness of a narrow strip of the planet during each revolution of the spacecraft, enables astronomers to assemble a composite image of Jupiter.

Bus—Electronics and Propulsion

Radiation Measuring Devices

Communication Antennas

Particle and Field Detectors

Plutonium Generators

Cameras and Imaging Devices

Using gyroscopes and celestial navigation, *Voyager 2* can point its instruments at specific targets while keeping a fixed orientation in space. Television-like cameras share the science boom with spectrometers and particle detectors; other instruments include a magnetometer and a whip antenna used to detect radio signals from planetary magnetic fields.

Sifting a Sea of Particles and Waves

Even when it is not taking pictures of planets, an exploratory spacecraft is hard at work analyzing the contents of interplanetary space. The apparent void is in fact laced with magnetic fields and bits of matter ranging from dust to atomic particles. The dust is believed to originate in the Solar System as debris from asteroids and comets. A thin, electrically charged gas, or plasma, streams from the Sun; this so-called solar wind is augmented by direct bursts of matter ejected by flares on the Sun's surface. From interstellar space comes a flow of cosmic rays composed of the nuclei of many kinds of atoms, some moving at nearly the speed of light.

Studying this interplanetary matter calls for a diverse collection of detectors, some of which are shown here, to record the density, energy, and composition of the particles. Space probes also carry magnetometers to measure the strength and orientation of the magnetic fields that surround the Sun and each of the planets. These fields interact with the particles of the solar wind and cosmic rays, which may change course or accelerate to high speeds near a planet.

Voyager's high-energy telescope (HET) is one of three kinds of particle detectors on the spacecraft; the others monitor low-energy cosmic rays and electrons. The tube-shaped HET permits entry of positively charged atomic nuclei—high-energy cosmic rays—through each end. Inside the tube is a stack of electrically charged plates. As a nucleus penetrates a plate, it causes a current to flow; the plates, in turn, slow the particles. By measuring the current produced in each plate and the number of plates the particle gets through, scientists can deduce the kind of atom the nucleus comes from, its direction, and its energy.

A meteoroid detector panel on a Pioneer probe can register the presence of a dust grain with a mass of a billionth of a gram. The detector incorporates eighteen gas-filled cells, each with a pair of electrodes at one end. Because the gas is pressurized, it cannot conduct a current between the electrodes, but when a tube is punctured by a dust particle, the resulting drop in pressure allows a current to flow. When the pressure drops still further, the gas again becomes nonconductive, stopping the current. The duration of the current flow indicates the size of the hole, and thus of the particle. Thirteen panels are attached to the back of the ship's main communications antenna.

Galileo's plasma instrument, shaped like a wedge of melon rind, is separated into two hollow shells by a charged middle plate, which bends the paths of entering particles: Electrons are attracted *(black tracks)*; positive ions are repulsed *(yellow)*. In one shell, electrons curve toward the outer wall while positive ions travel to the bottom of the detector; in the other compartment, the roles are reversed. Sensors measure the intensity of particles passing through each shell, and three mass spectrometers at the bottom of the inner one analyze some of the positive ions. A magnetic field in each spectrometer bends the path of a speeding ion; from the degree of bending, scientists can deduce the ion's chemical identity.

The Solar System is awash with charged particles. The Sun ejects mostly protons and electrons, which interact strongly with planetary magnetic fields. Cosmic radiation, a high-energy stream of atomic nuclei, enters from interstellar space.

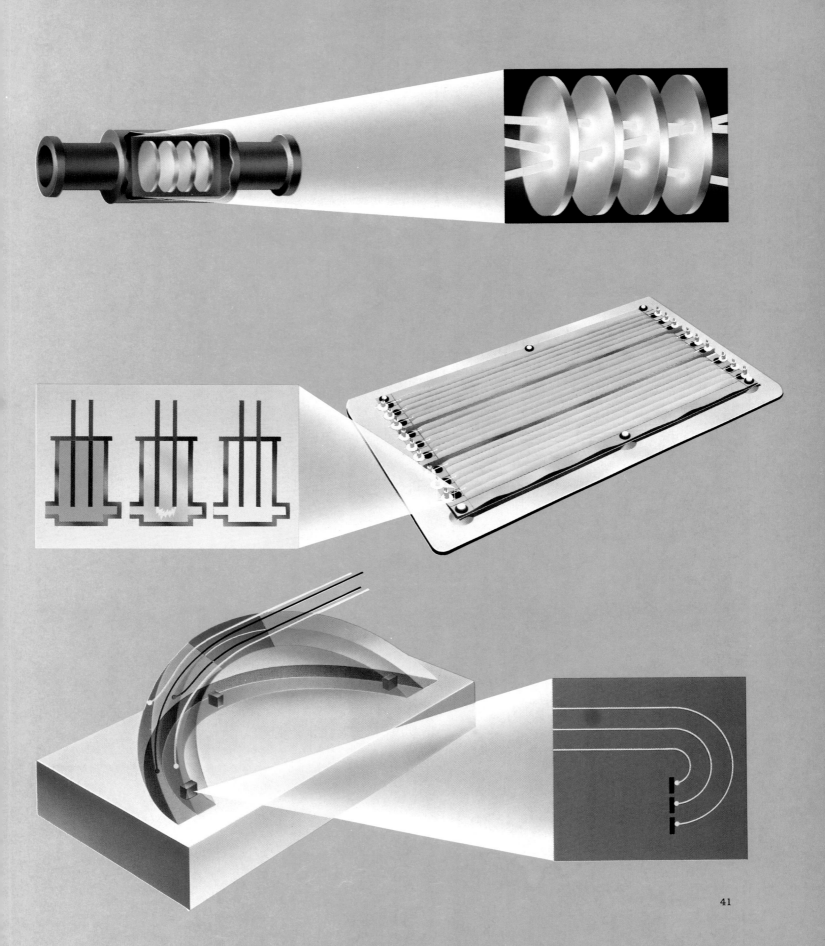

PICTURING
THE PLANETS

This *Voyager 1* image of Saturn, taken during its approach to the planet, is a typical unprocessed frame, exposed through a colored filter.

Three monochrome frames, colored with the blue, green, and orange of their respective filters, produce a fuzzy composite image.

The spellbinding pictures that herald the success of interplanetary probes are far more than simple snapshots: They are produced by an intricate system that has been enormously refined with each new generation of spacecraft. However, the basic principles of the picture-making process are common to all the probes. An electronic camera takes several black-and-white shots of the same subject, using a different color filter each time. Relayed to Earth, the data from these monochrome pictures is combined to produce one image in full color.

The highly successful Voyager probes use small television-type cameras. A single black-and-white image from one of these instruments is made up of 800 horizontal lines. For each line the camera records the brightness of 800 tiny dots, or pixels, each assigned a value ranging from 0 for black to 255 for white. These numbers are converted to a binary code (a string of zeros and ones) for transmission to Earth, where it is entered into a computer programmed to compensate for the slight differences in angle and distance that result from the spacecraft's movement between exposures. The individual images are colored according to the filters used and then combined to produce a full-color picture. Computer control gives astronomers the power to extract more information from the pictures by strengthening one hue relative to another, increasing contrast to bring out subtle features, or assigning arbitrary colors that highlight specific characteristics. The sequence below illustrates the image-making process, from the first monochrome picture to a false-color representation of atmospheric features.

Corrected to compensate for camera movement between shots, the monochrome frames are merged into a clear color picture.

The same image, enhanced by computer, highlights the low-contrast features of swirling clouds in Saturn's northern hemisphere.

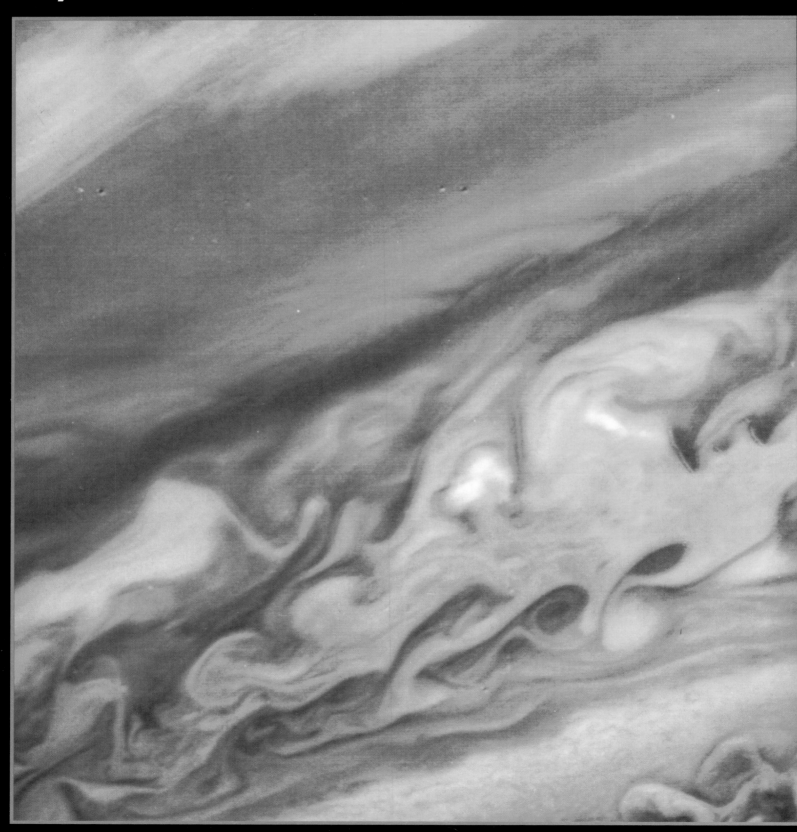

miles up is dominated by the Great
Red Spot, a centuries-old storm
three times the diameter of Earth.

uring 1903, Vesto Slipher at the Lowell Observatory in Flagstaff began taking observations of the outer planets with a new spectrograph that Percival Lowell had recently added to the observatory's arsenal. These devices, which divide incoming light into a rainbowlike fan of wavelengths, had been used by astronomers since the early nineteenth century. Distinctive patterns in the rainbows, the consequence of elements emitting or absorbing particular wavelengths *(pages 49-51)*, had revealed much about the chemistry and motion of stars. But obtaining spectra of planets, which mainly reflect energy received from the Sun, was more difficult and less informative. Jupiter was especially frustrating. The orange-red portion of its spectrum showed a set of dark smears that were impossible to decipher. The twenty-six-year-old Slipher, not long out of the University of Indiana, would go on to become a preeminent spectroscopist of stars and galaxies, but even in this early stage of his career, he was intent upon breaking new ground. In this instance, he aimed for the terra incognita of the near infrared, where visible light gives way to wavelengths too long to be perceived by the human eye. He applied experimental dyes and chemical baths to increase the sensitivity of his photographic plates to this unexplored region of the spectrum.

As he had hoped, patterns of absorption appeared, even for Jupiter. But they manifested themselves as broad bands of darkness rather than the fine lines seen in the spectra of stars, and he had no idea what elements caused them. His results, published in the *Lowell Observatory Bulletin,* became, as one contemporary astronomer put it, "one of the principal unsolved puzzles of spectroscopy." Many years later, in 1931, a quiet, aristocratic young chemist at the Georg August University in the medieval German town of Göttingen came upon Slipher's spectra and, with a stroke, cracked the code. In what colleagues later called "a brilliant piece of experimental and analytical work," Rupert Wildt realized that the bands were not from individual atoms, as astronomers had long assumed but could not prove. Instead, they were caused by more complex entities: molecules—specifically, methane and ammonia. Both molecules are rich in hydrogen, the most common element in the universe and the chief ingredient in stars.

Molecular absorption bands had never before been recognized in the spectra of a celestial object. Wildt had thus not only managed to accomplish an astronomical first, but he had done so with the most fugitive sort of evidence. The absorption patterns of methane and ammonia occur primarily in the far infrared, a region astronomers would not investigate for years. Discerning their traces in the near-visible wavelengths Slipher had probed was the astrochemical equivalent of identifying an elephant from a tuft of hair on its tail.

There were already considerable grounds for suspecting that the composition of the far planets differed from that of the rocky inner planets. For example, Jupiter's mass had been estimated centuries before by Isaac Newton from observations of the orbits of the Jovian moons. This mass, about a thousandth the figure for the Sun, yields an average density for the planet of about 1.3 grams per cubic centimeter, not much denser than water and roughly a fourth the average density of Earth. Given the spectroscopic evidence for the existence of large quantities of hydrogen on Jupiter (each methane molecule contains four atoms of hydrogen, each ammonia molecule, three), Wildt concluded that the giant planet, and perhaps the other far planets as well, was much more starlike than Earth-like in its makeup. Moreover, the fact that hydrogen was abundant on Jupiter but merely a trace in the terrestrial atmosphere shed light on how the Solar System had evolved. It tended to reinforce the general astronomical belief that the planets coalesced out of a cloud, or nebula, of material surrounding the Sun.

The process took many millions of years. First, dust and gas in the cloud gathered into small bodies called planetesimals, no more than a few miles across. The planetesimals in turn clumped together as they collided with one another, growing into full-size planets by about 4.6 billion years ago (a date estimated from the age of the oldest rocks on Earth). The low gravities of the less-massive planets could not hang on to volatile light elements like hydrogen, which boiled off into space. But the massive planets not only retained the hydrogen present at their creation, they also drew hydrogen from the congealing solar nebula into their deep, dense atmospheres.

The impact of Wildt's work was broad and immediate. According to the American spectroscopist Arthur Adel, who went on with Slipher to identify methane and ammonia as the sources of some fifty other absorption lines for Jupiter, Saturn, Uranus, and Neptune, the discovery "came as a bolt out of the blue." Recalling the work half a century later, Adel said, "The spectra obtained by Slipher lay around for generations, so Wildt's was a very substantial leap, a real shocker, especially when you consider that he was a chemist, not an astronomer."

Wildt emigrated to the United States in 1935, taking an appointment at California's Mount Wilson Observatory for a year, followed by five years at Princeton and four at the University of Virginia. During this period, the reserved chemist-turned-astrophysicist, like an illustrator adding detail to a sketch, continued to refine his conception of the far planets' atmospheres and deep interiors. In 1938, while at Princeton, he pointed out the importance to those models of the existence of metallic hydrogen—a form of hydrogen capable of conducting electrical current *(pages 70-71)*. In 1947 he proposed that hydrogen was the dominant element in Jupiter's atmosphere. The following year he moved to Yale University, where he would finish his career. Wildt developed the first realistic models for the interior of Jupiter: a solid core surrounded by a layer of compressed water and ice 20,000 miles thick, with an outer layer some 6,500 miles deep composed of compressed hydrogen, helium, and other gases. Four decades later, data from space probes would modify these proportions, but his central idea would hold. As extended by other astronomers since, it still shapes perceptions of the world beneath a raging, many-colored atmosphere.

JUPITER SPEAKS

Rupert Wildt was one of a handful of original thinkers, armed with insight, doggedness, and luck, to take planetary studies seriously in the years preceding the epoch of interplanetary flight. Similar traits could be seen in the efforts of two scientists working in the very different field of radio astronomy—and the payoff was equally remarkable.

On a winter day in 1955, Bernard Burke and Kenneth Franklin stood in a fallow Maryland field near Washington, D.C., preparing to survey a small section of what had become known as the radio sky—the celestial outpouring of radio energy by stars and galaxies. Burke, a physicist with the Carnegie Institution's Department of Terrestrial Magnetism, and Franklin, his astronomer assistant, had been mapping the sky for the past few months, using a cross-shaped antenna more than a thousand feet on a side, tuned to receive radio waves only at a frequency of 22.2 megahertz, or million cycles per second. Radio waves at this frequency are about thirteen meters long, longer than wavelengths astronomers had used until these two scientists set up their huge antenna to, as Franklin put it later, "see what was there."

Occasionally, a burst of radio energy from an unknown source would trigger the stylus of the antenna recorder. Since the strongest known radio source in their antenna's field of view was the Crab nebula, the fuzzy remains of an ancient supernova explosion, they assumed the rogue radio burst came from some common, earthly generator—perhaps the ignition of a car or tractor. They ignored it. Their immediate concern was whether to orient the tangled nest of cables and wires that constituted their antenna toward the north or

Spanning the visible spectrum from violet to red, a representation of a spectrogram of the Sun's light *(below)* reveals an ordered array of some of the more prominent absorption lines— among them, a narrowly spaced pair just below 5,900 angstroms indicating the presence of sodium.

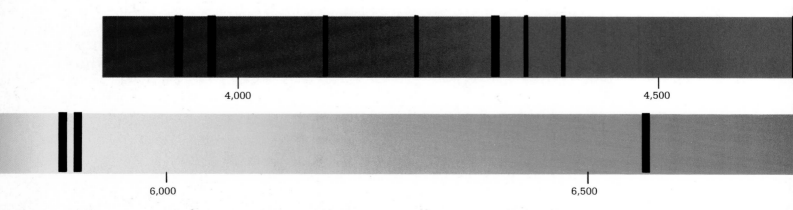

4,000   4,500

6,000   6,500

THE HIDDEN MESSAGE OF LIGHT

When the French philosopher Auguste Comte declared, early in the nineteenth century, that the distant celestial fires would remain forever unknowable, he did not bargain on the information that would soon be discovered within light itself.

Isaac Newton had already used a prism to show that sunlight is made up of a rainbow of colors ranging from red to violet. This solar spectrum was later seen to be striped by hundreds of dark lines. Their meaning began to come clear in the 1850s, when German scientists Robert Bunsen and Gustav Kirchhoff burned different substances in the laboratory and found that the result was a series of bright lines of color in characteristic patterns rather than a continuous spectrum.

By 1859 Kirchhoff had compared the bright lines given off by elements such as hydrogen and iron with certain dark lines in the solar spectrum and discovered that the patterns matched exactly. He concluded that the dark lines were the thumbprint of elements in the Sun—gases in the solar atmosphere that absorbed particular wavelengths from the light passing through them. (The wavelengths are measured in units called angstroms—equivalent to .000000004 inch.)

Today, astronomers studying celestial objects can identify a vast number of atomic and molecular species by their spectral signatures. To do this, they use a machine called a spectrograph. Typically, such a device will employ a diffraction grating—a mirror fashioned into a miniature staircase with thousands of tiny steps—to break light into its constituent colors. The results may be displayed as a diagram, in which upward spikes represent emission lines and downward spikes denote absorption lines, or they may be recorded as a photographic image, with bright and dark lines appearing much as they did to Kirchhoff more than a century ago.

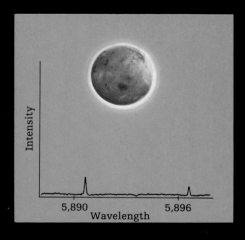

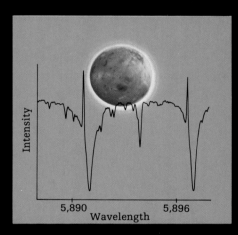

The two highest peaks in the spectrogram shown at far left signal the presence of evaporated sodium in the atmosphere of Jupiter's moon Io. The emission lines are produced as sodium atoms absorb and then reradiate the Sun's light at specific wavelengths.

At near left is a spectrogram of the moon itself. The emission lines from Io's atmospheric sodium are superimposed on a spectrum of sunlight reflected from the moon's surface. Deep sodium absorption lines in the solar spectrum show that the element is present in the solar atmosphere as well.

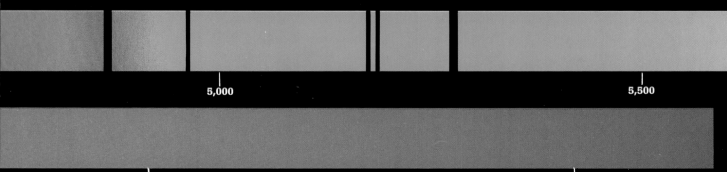

5,000   5,500

7,000   7,500

An Interplay of Energies

Spectral lines are born when light interacts with gas—specifically, when packets of light energy, or photons, are gobbled up or released by the atoms that make up a gas. A common example is the pattern of lines produced by the interaction of light and hydrogen, represented here in a simplified model.

A hydrogen atom consists of a single electron orbiting a single proton. That electron can exist in any of a series of discrete orbits, each of which has a particular energy level associated with it. Six levels are shown here. Electrons orbiting far from the atomic nucleus contain more energy than those near it, and the energy of the atom as a whole depends on the orbit of its electron at a given instant.

A photon of light penetrating the hydrogen atom also has a specific amount of energy, depending on its wavelength. Photons of short-wavelength blue light, for example, contain more than those of long-wavelength red light.

Given a suitable injection of energy, the hydrogen electron will leap from a low orbit to a higher one. The additional power has to be supplied by a photon carrying precisely the required amount: that is, one with a very specific wavelength. In the process, the electron in the atom absorbs the photon, and this action, repeated in countless atoms, creates an absorption line in the visible spectrum that reaches the astronomer.

The reverse occurs when a gas is losing, rather than gaining, energy—say, when a warm hydrogen cloud cools down in a cold environment. In this case, electrons will drop to lower energy levels and release energy in the form of photons, thereby producing emission lines in the spectrum.

Complicating the astronomer's task are additional complex spectral line patterns, produced by the action of molecules spinning and vibrating. Yet these lines, too, reveal unique information about the gas's composition, temperature, and density.

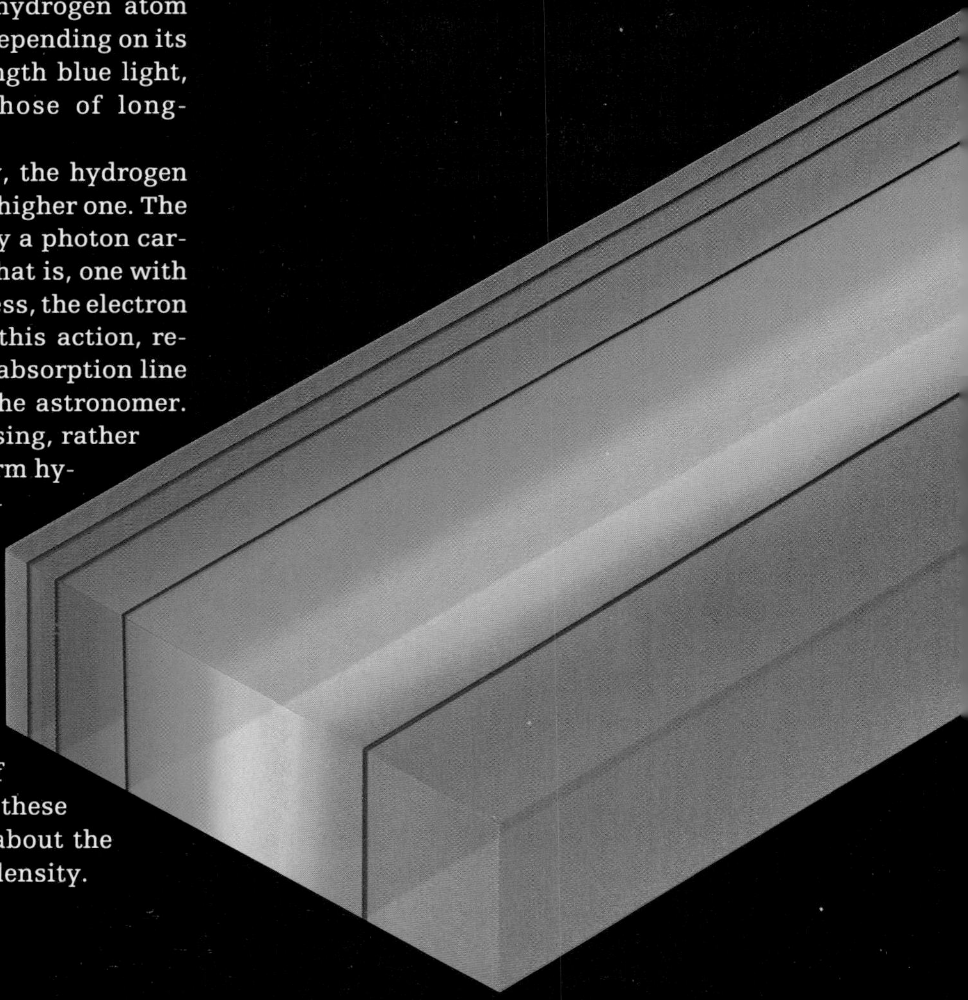

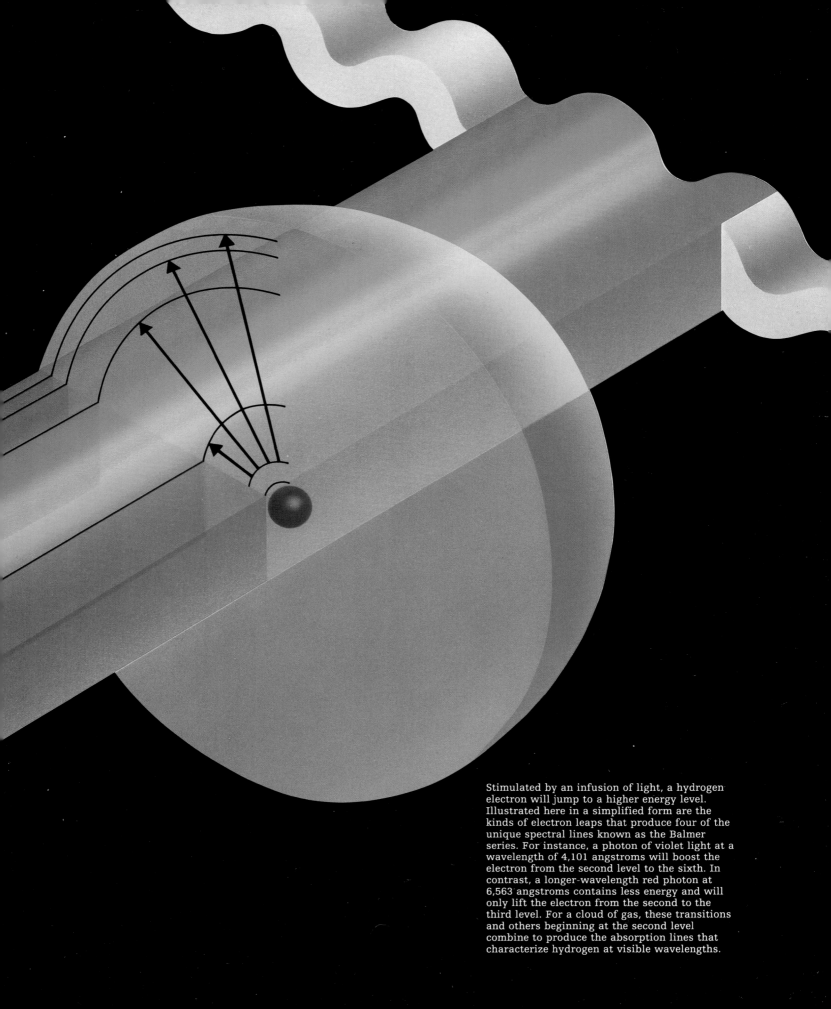

Stimulated by an infusion of light, a hydrogen electron will jump to a higher energy level. Illustrated here in a simplified form are the kinds of electron leaps that produce four of the unique spectral lines known as the Balmer series. For instance, a photon of violet light at a wavelength of 4,101 angstroms will boost the electron from the second level to the sixth. In contrast, a longer-wavelength red photon at 6,563 angstroms contains less energy and will only lift the electron from the second to the third level. For a cloud of gas, these transitions and others beginning at the second level combine to produce the absorption lines that characterize hydrogen at visible wavelengths.

the south for the next series of observations. Years afterward Franklin would point to this moment as a prime example of the way serendipity operates in science. Had they aimed their antenna north, the mysterious interference would have faded and finally died. Instead, they turned southward and unwittingly tracked the source across the sky.

Many weeks later, on a Monday morning in late March, Burke summoned Franklin into a large room shared by the department staff at the Carnegie Institution's offices in Washington. Three months' worth of strip charts were draped across a drafting table, each scored with the squiggly signatures of incoming radio waves. The pattern was obvious: Each nightly trace showed a peak representing the radio signal from the Crab nebula and another, more pronounced peak representing the interfering radio signal. No earthbound source was likely to harmonize with the Crab nebula night after night. Something unknown ghosted along, now and then muttering radio energy.

At about this point, Franklin recalled much later, a researcher named Howard Tatel suggested that the source might be Jupiter. Franklin's first reaction was, "Oh boy, that's ridiculous." Nevertheless, he fetched a copy of the *American Ephemeris and Nautical Almanac,* which lists the predicted positions of the Sun, Moon, and planets on the sky, and looked at the celestial coordinates of Jupiter. Sure enough, for the past few months, the great planet had been sailing through the region they had just surveyed. Franklin decided to find out if Tatel was right. As soon as he reached his desk at Carnegie the next day, he began graphing the coordinates of the mysterious source against Jupiter's position. "Bernie came in and was looking over my shoulder," Franklin said later. "Every time I plotted a point he said, 'Wow!' " The match between the two sets of coordinates was compellingly close. "This was Tuesday morning. By Tuesday afternoon, the whole lab was buzzing." Jupiter, a planet, was emitting radio energy like a star.

Soon tens of thousands of people would literally hear the voice of Jove. At the request of the National Broadcasting Company, Franklin hooked a tape recorder up to the speaker on the radio antenna—a speaker used mainly to listen for interference—and recorded the swishing static emitted by the giant planet. NBC aired the tape on their first "Monitor" program, a pioneering weekend radio magazine series. Remembers Franklin, "We came right after noises from oysters."

THE DYNAMO

Jupiter may have spoken, but in a language at first too arcane for scientists on Earth to understand. The radio emissions arrived in sporadic bursts that lasted sometimes five minutes, sometimes two hours, and carried incredible amounts of energy—on the order of 10,000 megawatts, or about the equivalent of a billion simultaneous lightning strikes. Astronomers were at a loss even to hypothesize a source for such power.

The Jovian message was complicated several years later when scientists at the Naval Research Laboratory in Washington, D.C., using a new eighty-

Good land-based photographs of Jupiter, including this one, which was taken through an eighty-two-inch reflector telescope, are sometimes more useful than those from the Voyager craft. Because earthbound astronomers can observe the planet continuously for years, they are able to record day-to-day changes in atmospheric features such as "festoons," scroll-like structures here visible in Jupiter's central white zone.

four-foot radio telescope erected south of the city, discovered that Jupiter also gave off signals at wavelengths in the centimeter range, hundreds of times shorter than the radio waves detected by Burke and Franklin. According to thermodynamic law, any perfect absorber and radiator of heat—a "black body"—emits its peak energy at progressively shorter wavelengths as the body's temperature increases. Thus, an object producing mainly radio energy at relatively low temperatures should shift to infrared wavelengths as it warms up, then visible light, and finally to high-energy ultraviolet, x-ray, and gamma radiation as its temperature continues to rise.

Early in 1958, the Naval Research Laboratory staff compared Jupiter's short-wavelength energy output to that of a black body and derived the planet's so-called brightness temperature—equivalent to an object's real temperature, provided the radiation comes from a thermal source. At a three-centimeter wavelength, the brightness temperature was about 150 degrees Kelvin. Jupiter had been observed earlier in the near infrared, and those measurements had also yielded a reading of about 150 degrees Kelvin. Since infrared light comes from emitted heat, this was considered to be close to the actual temperature of Jupiter's cloud tops. But at a wavelength of ten centimeters, the brightness temperature, instead of dropping, jumped to 600 degrees, a value that dramatically violated thermodynamic law.

In the summer of 1958, the Naval Research Laboratory's Edward McClain described this puzzling observation at a symposium in Paris. His talk drew the attention of Frank Drake, a young astronomer at the National Radio Astronomy Observatory in Green Bank, West Virginia. Troubled by the difference between what had been observed and what was thermodynamically possible, Drake and a collaborator painstakingly gathered radio data from Jupiter at longer centimeter wavelengths, beginning in March 1959. To sharpen their reception, the astronomers called for the radio dish to be moved every thirty seconds for hours on end, first toward, then away from, the target. Since the dish was guided manually, Drake told colleagues later, having to make the adjustments "drove the telescope operators crazy."

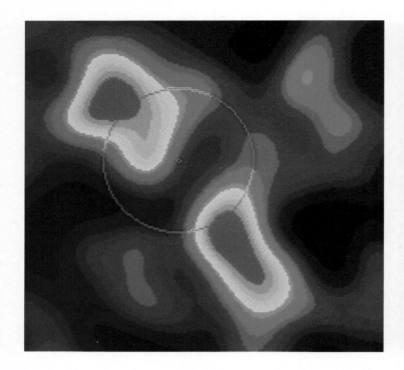
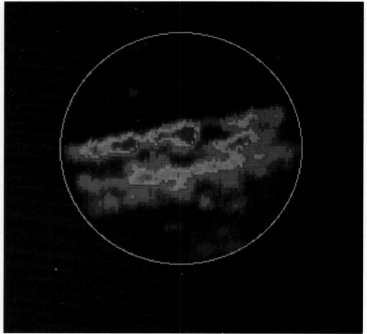

But the technique paid off. At a wavelength of twenty-two centimeters, Jupiter's brightness temperature was 3,000 degrees Kelvin, or about 5,400 degrees Fahrenheit—scarcely 1,500 degrees cooler than the surface of the Sun. Convinced by this improbable value that Jupiter's radiation must come from a nonthermal source, Drake systematically increased his listening wavelength. At a wavelength of sixty-eight centimeters, Jupiter's brightness temperature leaped to 70,000 degrees Kelvin, or about 90,000 degrees Fahrenheit, hotter than the surfaces of the hottest stars.

Through complex mathematical analysis, Drake established that the short-wavelength radio emissions from Jupiter resembled those from such distant radio sources as the Crab nebula. He concluded that the enormous energies seen at Jupiter were generated by electrons moving through a magnetic field at nearly the speed of light. Called synchrotron radiation because it was first observed in the early particle accelerators of the 1940s, the process was a well-known stellar phenomenon but had only recently been identified as a planetary possibility as well. The year before, in an experiment directed by James Van Allen of Iowa State University, sensors on the U.S. Army satellite *Explorer I* had detected Earth-girdling belts of radiation (later named Van Allen belts) formed by subatomic particles trapped in Earth's magnetic field.

The particles—electrons and electrically charged hydrogen and helium atoms—escape from the Sun's upper atmosphere, or corona, at a rate of a million tons per second and flow out into space in what is called the solar wind. Gusting to velocities of 600 miles per second or more, this wind has an average speed past Earth of about 250 miles per second. Interactions between the solar wind and magnetic field's lines of force sculpt the field into a long, windsock-shaped cocoon called the magnetosphere. In these interactions, some particles become trapped in the field and are concentrated into belts of intense radiation. Drake proposed that similar belts might exist at Jupiter.

Within a few months, astronomers at the California Institute of Technology

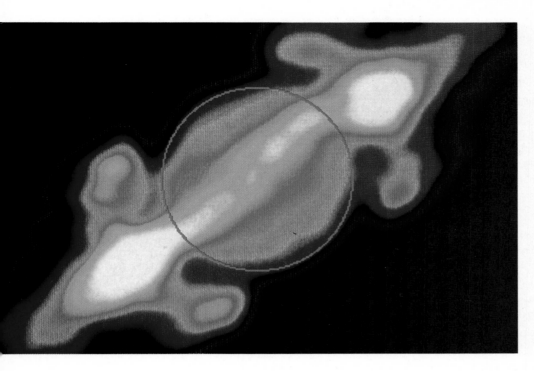

Jupiter radiates energy at a variety of wavelengths, yielding strikingly different images and kinds of information at different places on the spectrum, as shown by the images at left. (In each case, Jupiter's disk is outlined in red.) For instance, an x-ray portrait of the planet *(far left)* reveals aurorae streaming out along its magnetic poles. The x-rays, produced by the collision of electrons with Jupiter's atmosphere, are most intense in the white areas of the image.

At infrared wavelengths *(center)*, heat from Jupiter's interior is detected escaping primarily along two bands at the planet's equator. The hottest zones show up as orange and red. Infrared measurements reveal that deeper in the giant's atmosphere temperatures rise dramatically.

In a radio map *(near left)*, waves produced by electrons spiraling within Jupiter's magnetic field emanate from belts at the planet's magnetic equator. White regions indicate greatest intensity.

confirmed Drake's hypothesis. Their radio telescopes delineated a ring of charged particles around Jupiter that rotated with the planet and stretched past the moons Io and Europa to a distance three times Jupiter's diameter. Scientists had their first intimations of Jupiter's vast magnetosphere.

Despite the progress on the short-wavelength front, the source of the long-wavelength emissions recorded by Burke and Franklin remained in dispute. A front-runner among several competing explanations was proposed by University of Colorado astrophysicist James Warwick in 1960. In Warwick's model, Jupiter's magnetic field was not centered on the planet's own core but on a point closer to the Jovian surface. This would cause particles trapped within the field to move eccentrically and bunch up, sending out an electromagnetic shock wave in the form of a burst of radio energy.

The idea proved wrong, but Warwick's work ultimately led to the correct answer. Over the course of four years, he had accumulated and cataloged data on Jupiter's long-wave radio emissions. In 1964 he loaned his catalogs to visiting Australian researcher E. Keith Bigg. The scientist from Down Under was interested in supposed lunar effects on terrestrial rainfall and was studying the moons of Jupiter as a possible analogue. Just two days later, Warwick recalled, the Aussie bounded back in. "What he had was gorgeous."

Bigg had shown unequivocally that the intensity of the long-wave radio bursts received at Earth varied according to Io's position in its orbit. Only when Io crossed Jupiter's magnetic equator on one side did the sporadic emissions occur. The tiny moon, flying through the planet's powerful magnetic field, acted like the brushes in a gigantic dynamo, producing Jupiter's guttural hiss of long-wave radio energy.

SCOUTS OUT
Within a decade of Bigg's discovery, Io would be joined briefly by another moon—this one artificial. Space had become the province of manned flight

and of a new science that used robot surrogates as probes to other worlds. An electronic voice would join Jupiter's, returning a trill of numbers in which scientists would see the distant giant clearly for the first time.

Spacecraft exploration of the far planets began in March 1972 at Cape Canaveral, Florida, when the first round in a volley of powerful rockets launched a 550-pound spacecraft called *Pioneer 10*, descendant of a series of solar-wind probes. Accelerating to velocities that broke the thread of terrestrial gravity, *Pioneer 10* began its twenty-month-long glide toward Jupiter, blazing the first trail to the planets beyond Mars. A year later a second probe, *Pioneer 11*, rose from Florida and followed in its wake.

By December 1973 *Pioneer 10*'s first transmissions of the Jupiter encounter were chattering into mission control at the National Aeronautics and Space Administration's (NASA's) Ames Research Center. Located south of San Francisco, Ames was where the Pioneer probes had been developed. The rising digital chorus from this pathbreaking robot revealed that Jupiter, while defined in a general way from the ground, was in every respect designed on a grander scale than scientists had thought.

The planet sailed the solar wind like a great ship, surrounded by a powerful magnetic field blunted at its sunward end into a bow wave and tailing off for hundreds of millions of miles downwind. Within that magnetic pouch, flattened bands of high-energy particles spawned radiation ten thousand to a million times more intense than that found in Earth's Van Allen belts.

As the spacecraft fell through this vast cloud, its systems reeled from the shocks of gamma and x-rays generated by the charged particles, so numerous that they saturated the probe's particle counters. A dazed *Pioneer 10* muttered spurious commands to its instrumentation, causing the loss of several planned high-resolution images of Jupiter and its moons. And then, as anxious mission scientists watched from Ames, the radiation levels flattened out. A bit scorched and sickened, *Pioneer 10* had survived the first crossing.

This resilient little craft's brief, dangerous encounter with Jupiter fundamentally altered the human view of the Solar System's largest player and its coterie of moons. Among other things, the sensors aboard *Pioneer 10* confirmed Rupert Wildt's forty-year-old hypothesis that Jupiter was not like the inner planets in its composition but something closer to a star. Its core, heated by the gravitational contractions of the still-coalescing planet, was a powerhouse in its own right. Like a cooling coal, Jupiter radiated energy—about twice as much as it received from the Sun.

Almost exactly one year later, as *Pioneer 10* glided off toward the edge of the Solar System, its job done, the next *Pioneer* swept past Jupiter, adding its numerical footnotes to the data taken by its twin. This craft was not merely scanning Jupiter, it was also using it. The trajectory of *Pioneer 11* had been designed so that Jupiter's gravity would bend it, whipping the tiny craft across the reaches of space in pursuit of Saturn.

The *Pioneers* were also in pursuit of a credential. Both craft had been conceived in the early days of the space age, their missions built around limited resources. Justifiably proud of their machines, the Pioneer scientists at Ames were thus a bit defensive about them as well. One went so far as to argue that later probes added nothing to what the *Pioneers* had found, but the argument had the distinct flavor of sour grapes. These first two spacecraft were scouts, and like scouts, they traveled light. Their imaging systems were dim-sighted and slow by later standards, their instruments designed to do little more than measure field strengths and energy levels. A fundamental purpose of their missions was simply to see if space probes could survive intense radiation and the hazards of the asteroid belt, where thousands of rocky chunks ranging in size from 5 to 500 miles are scattered like islands in a loose-knit, drifting archipelago.

The *Pioneers* broke their trails splendidly. But even before the first of them roared spaceward, scientists at NASA and the Jet Propulsion Laboratory, or JPL, were planning an ambitious outer-planet odyssey that would give scientists their first truly intimate view of Jupiter and its fellow giants.

NOW VOYAGER
In early 1972 NASA had solicited proposals for experiments to fly aboard the planned successor to the *Pioneers*—a spacecraft more than three times as heavy and many times more capable than the earlier probes. The mission grew out of a scheme called the Outer Planets Grand Tour Project, designed to take advantage of a rare event. Once every 177 years—give or take a year—Jupiter, Saturn, Uranus, and Neptune swing into a rough kind of line as viewed from Earth. The last time they had been thus positioned was during the era of Jefferson, Napoleon, and George III. The arrangement does not last long; for NASA, the critical period, from 1976 through 1978, was rapidly approaching.

During this brief interval, a spacecraft launched from Earth could use the gravity at each successive planet both to bend its trajectory toward the next destination and to gain speed in the process. Without such a gravity assist, long interplanetary journeys of exploration were beyond existing technology. With it, two probes could be launched, one toward a rendezvous with Jupiter, Saturn, and Pluto, the other toward Jupiter, Uranus, and Neptune. But the budgetary realities of the 1960s soon cooked the Grand Tour venture down to something the 1970s could afford: a mission with the working title of Mariner/Jupiter-Saturn, or MJS. Three craft would be built. Two of them would be launched for encounters with the two giant planets. One would be held back as a spare and possibly would end its days in the National Air and Space Museum in Washington, D.C.

The initial $250 million program took form around an 1,800-pound spacecraft that was both more versatile than its predecessors and uncommonly independent. Instead of merely noting what passed before its sensors, the probe could aim its instruments at specific targets, making it an ideal camera platform. Most important, it carried three interconnected pairs of computers that were capable of dealing with catastrophes ranging from stuck fuel valves to malfunctioning gyroscopes, without asking for help from Earth. Once the craft reached Jupiter, it would be forty minutes away by radio, and a lot could go wrong while it waited for instructions.

The probe's computers were taught an elaborate language to describe what the video cameras saw. Each image would be made up of 640,000 picture elements, or pixels—the equivalent of dots in a newspaper photograph. The individual pixels, produced by scanning an image that formed on an electrically charged plate at the back of the camera, could be assigned a brightness level ranging from 0 for pitch black to 255 for white. The probe's computers would transmit these brightness levels in a binary code of zeros and ones—eight bits, or binary digits, per pixel, amounting to a five-million-bit description for each of the thousands of images to be returned from Jupiter.

To accommodate this rich numerical prose style, mission designers installed radios tuned to the so-called X-band at 8.4 gigahertz (billion cycles per second), a radio frequency high enough to move the mountain of data in short transmitting times. As the actual radios began to arrive at JPL, however, mission managers discovered that only seven in a batch of fifty-two built for the project worked reliably. Worried that their probes might be mute beyond Jupiter, they decided to include a simpler system that would return imaging data on the S-band frequency, at 2.3 gigahertz. Long the workhorse frequency of space exploration, the S-band had been selected for the two-way communication of both science and "housekeeping" messages. If it were used to transmit images, mission controllers would have to be patient: The lower frequency meant fewer bits of transmitted data per unit time. But the demonstrated reliability of S-band radio would guarantee an articulate probe at Jupiter and Saturn if the temperamental X-band units failed.

In every respect, the Mariner/Jupiter-Saturn vehicle was the finest flower of planetary exploration, alongside which the *Pioneers* seemed positively dowdy. By 1977 the new spacecraft had everything but a proper name. Leading candidates were Pilgrim, Nomad, Discovery, Voyager, and names, such as Orion, taken from the stars. Finally, NASA and JPL chose Voyager, although with some misgivings: An earlier project by that name had been canceled.

For a time, it must have seemed to the Voyager team that their superstition had been well-founded. From California to Kennedy Space Center, from Cape Canaveral to the asteroid belt, the mission became an unnerving litany of glitches. The spacecraft designated *Voyager 2* was scheduled for first launch because its trajectory, modified to include an optional flyby of Uranus after Saturn, would put it at Jupiter after the second-launched probe, designated *Voyager 1.* But the original *Voyager 2* developed mysterious disorders of the control and flight data systems and had to be swapped, along with some equipment, with its twin.

On August 20, 1977, *Voyager 2* blasted off from Cape Canaveral and almost immediately sent back a message of fresh troubles. To begin with, the craft's onboard computers presented so many problems that they seemed, according to one observer, "almost humanly perverse—and perhaps a little psychotic." Not only did the electronic brains sometimes refuse to carry out ground controllers' commands, they also issued random orders of their own, firing the thrusters at whim and miscuing other vital instruments. Later, controllers realized that *Voyager 2*'s computers, surprised by the rapid acceleration at launch, thought they had an emergency and embarked on their own built-in programming sequence to rescue the mission.

Then mechanical gremlins went to work. Less than an hour into the flight, the boom carrying most of the scientific gear appeared to jam as it unfolded. Several days' nervous fiddling by ground controllers indicated that the problem was not with the boom but with the sensor monitoring its position.

Equally worrisome was a looming fuel shortage. A solid-propellant kick rocket used to add a final burst of speed had separated from *Voyager 2* as planned, but its supporting struts, which remained with the spacecraft, partially blocked the small rocket thrusters used to control the probe's trajectory. The resulting loss of thruster efficiency cost nearly 15 percent more fuel than intended; unless something was done, the thrusters would run dry beyond Saturn and foreclose the optional Uranus mission. The solution: Critical trajectory-correction maneuvers, initially meant to be executed after the Jupiter encounter, were rescheduled to take place during the encounter period itself. The craft's added speed near the planet could thus be tapped for some of the course-changing energy, conserving precious fuel. These changes, which effectively redesigned the mission with the birds in flight, set the imaginative, innovative tone of the entire Voyager odyssey, as creative technicians on the ground altered both the mission and the spacecraft over distances of millions, then billions, of miles.

Scheduled for launch only twelve days after *Voyager 2*, *Voyager 1* waited at Cape Canaveral while mission controllers monitored the disheartening drama unfolding around the flight of its sibling. Then, after a four-day delay and some adaptations learned from the first-launched craft, *Voyager 1* lifted off on September 5.

Even as its twin sped in pursuit, *Voyager 2* played yet another wild card. Eight months into its flight, the craft's two S-band radio re-

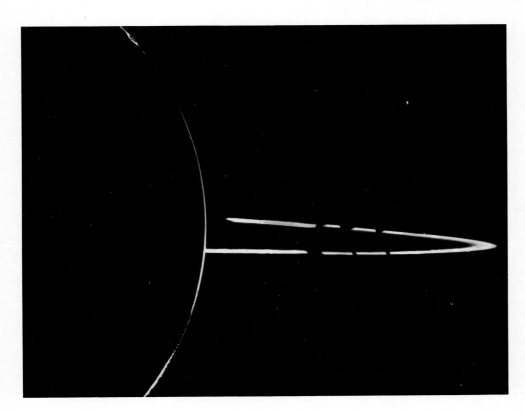

A mosaic of images from *Voyager 2* shows Jupiter's delicate rings gleaming in the sunlight. So thin as to be virtually invisible from Earth, the bands, which extend about 32,000 miles out from the planet's atmosphere, were first discovered by *Voyager 1* in 1979.

ceivers caught the mission's chronic malaise: The primary radio receiver blew a fuse, and the backup, with malfunctioning electronics, would not track incoming signals. Again, patient controllers sorted through the possibilities. They discovered that *Voyager 2* had suffered a kind of hearing loss. Where before it had been able to receive signals up to 100,000 hertz (cycles per second) on either side of its center frequency—the S-band's 2.3 gigahertz—it now would accept only signals within 96 hertz. This posed severe problems, since the radio's center frequency drifted with the shifting orientation of the spacecraft to the Sun and with slight changes in temperature caused by the switching on and off of electronic components. Controllers would have to aim their signals very close to a moving frequency mark if they wanted to talk to *Voyager 2.*

After a few tense days of chasing the probe's listening frequencies, they found enough pattern in the seemingly random tuning to develop an elaborate computer program that, run on the big machines at JPL, could predict *Voyager 2's* center frequency. Now the engineers were able to get back on the air to the spacecraft's computers and resume their interrupted dialogue. Although the drifting frequency of *Voyager 2's* radios would plague the mission from this point onward, the little ship was finally rigged for Jupiter.

In mid-December, both *Voyagers* were nearly 80 million miles from Earth and 11 million miles apart, with *Voyager 1* beginning to pull ahead. Just over a year into their journey, they were safely across the minefield of the asteroid belt. A bright, banded giant of a planet expanded in their fields of view.

GREAT JOVE

From the moment *Voyager 1* crossed the bow wave of the Jovian magnetosphere *(pages 76-77)* in February 1979, it began to tear away the veils of Jupiter. In Pasadena, the mission team, and the hundreds of journalists camped at JPL for the encounter, were treated to a nonstop circus of dazzling images. Monitors there filled with high-resolution pictures of Jupiter that captured the yellow and burnt-orange sphere with unprecedented clarity. The vague whorl of the Great Red Spot leaped into focus as a gigantic Jovian storm, and the atmosphere, crackling with lightning and aurorae, revealed eddies, wind bands, and hurricane-like spirals only hinted at by the cameras on the *Pioneers.* After watching the first kaleidoscopic images of the scalloped Jovian clouds flash across the lab's television screens, the University of Arizona's Bradford Smith, head of the Voyager imaging team, announced, "The existing circulation models have all been shot to hell. They fail entirely in explaining the detailed behavior that *Voyager* is now revealing." His comment, and others like it, became a kind of anthem for what followed.

As the two *Voyagers* took turns with Jupiter in the spring and summer of 1979, the rest of the planet's wonders grabbed center stage, including a diaphanous ring that had been the subject of speculation among planetary astronomers since the 1960s, when its existence had first been proposed.

No one expected Jupiter to be a ringworld comparable to Saturn, but *Pi-*

Orbiting about 260,000 miles above the cloud tops of Jupiter, Io appears deceptively placid to the *Voyager 2* spacecraft, eight million miles away. Two weeks after this picture was taken, the probe made its nearest approach to the Jovian moon, recording data that confirmed the volcanic activity that makes Io one of the most unusual moons in the Solar System.

REVELATIONS OF A RUSSET MOON

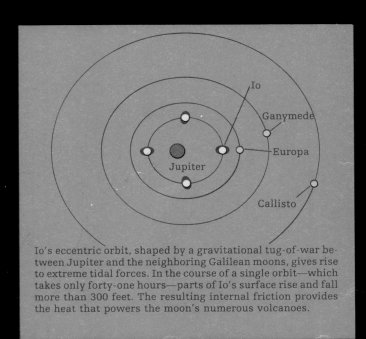

Io's eccentric orbit, shaped by a gravitational tug-of-war between Jupiter and the neighboring Galilean moons, gives rise to extreme tidal forces. In the course of a single orbit—which takes only forty-one hours—parts of Io's surface rise and fall more than 300 feet. The resulting internal friction provides the heat that powers the moon's numerous volcanoes.

Io, innermost of Jupiter's large moons, showed the Voyager probes a face unlike any seen in the Solar System. Plumes of gas and solid particles, ejected by enormous volcanoes, rose more than a hundred miles above a surface splotched with multicolored sulfur lava. Huge patches of frozen sulfur dioxide reflected the light of the Sun, making Io six times brighter than Earth's moon, which is the same size. Beneath the surface, scientists concluded, lay a seething caldron of molten sulfur and liquid sulfur dioxide.

Scientists had long suspected that Io's orange-red surface was caused by deposits of sulfur, but its intense volcanism was a surprise. The Voyager photos of Io's huge volcanic plumes were the first evidence of active volcanoes beyond Earth. The photos also showed that the moon's surface was completely free of impact craters, ordinarily common on planets and moons subject to regular meteor strikes. Relentless volcanic activity constantly renews much of Io's landscape, quickly covering craters with flowing lava and drifting debris from the plumes. Even in the brief four-month span between the two Voyager missions, some volcanoes ended their eruptions, leaving vast patterns of deposits that changed their regions almost beyond recognition.

BRIMSTONE ON A DISTANT WORLD

Unlike the high, crater-topped volcanoes found on Earth, Mars, and Earth's moon, Io's volcanoes are sunk deep into its surface. Their plumes issue from narrow fissures or from broad calderas—depressions formed by the collapse of subsurface chambers emptied of lava in previous eruptions. Gas and debris spew forth with the force of rifle bullets, rapidly mushrooming in Io's low gravity and thin atmosphere. The highest plumes observed by the Voyager probes extended 190 miles above the surface, with sulfurous fallout changing the color of areas as large as Alaska.

The volcanic eruptions are of two types. The largest plumes—usually higher than seventy-five miles—last for days or weeks. They apparently occur when Io's internal heat volatilizes subsurface sulfur lakes. The red and orange particles from these plumes fall back to the surface to give Io its characteristic rust color. The smaller eruptions, which may last for years, consist of sulfur dioxide gas that drifts back down as bright white sulfuric snow. Such snow covers 30 percent of the moon's surface.

Scientists who hope to use Io's spectacular geology to gain insight into volcanoes on Earth may be beneficiaries of a little cosmic luck. Many believe that the volcanic activity is periodic. The Voyager probes arrived during a period of violent eruptions, thought to last about 10 million years; the phase may be succeeded by 100 million years of calm.

The mottled, many-hued surface of Io bears witness to the moon's fiery nature. One prominent feature, the light-colored ring near the center, is a volcano that was erupting when this photo was taken by *Voyager 1,* from a distance of about 500,000 miles.

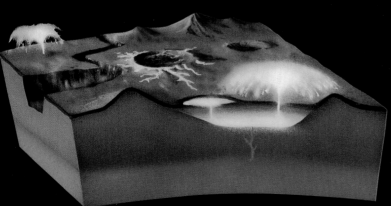

One model for Io's volcanism shows a thin crust of solid sulfur and sulfur dioxide covering a subcrust of silicate rock *(gray)*. Depressions in the subcrust cradle reservoirs of liquid sulfur *(yellow)*, heated by the molten core *(red)*. Liquid sulfur dioxide permeates the subcrust, escaping as vapor from the walls of crevices. Eruptions occur when liquid sulfur dioxide contacts hot sulfur and vaporizes explosively.

Meandering flows of
radiate from a dark v
caldera *(left),* one of I
volcanic hot spots. Th
steep-walled depres
from 12 to 125 mi
thought to cont
sulfur lakes a
basaltic lav

Layer upon layer of frozen
ide and multicolored form
create a painted desert
surface. The volca
so active that
years Io
ane

Taking leave of Jupiter and its moons, *Voyager 2* captures the traces of two volcanic eruptions on Io's crescent. First seen by *Voyager 1* four months earlier, these relatively small plumes arch more than sixty miles into the moon's thin sulfur dioxide atmosphere.

oneer 11 had indirectly detected the ring as a transient dip in the number of charged particles striking the spacecraft. On the chance they might spot something, JPL scientists programmed *Voyager 1* to take a single, long-exposure image where they guessed a ring might be. And it was: barely visible but definitely present. In later shots from *Voyager 2*, the ring, some 4,000 miles wide but no more than 20 miles thick, arcs across the blackness, backlit by the Sun. Cleaned up by computers to eliminate the smearing caused by long exposure times, the images revealed that what seemed to be a single band was actually three, gently linked to the tops of Jupiter's clouds by a thin veil.

ASTOUNDING, CONFOUNDING IO

Into this heady atmosphere of fresh discovery came the first shots of little Io, one of Jupiter's sixteen moons. "In the press room," recounted a science writer who was there, "reporters gaped at the monitors and tried to think of new adjectives." Scientists, whooping with joy and amazement, did not know what to make of the satellite. "It shouldn't look like this at all," cried one. Its mottled orange, yellow, and white impasto was compared to moldy oranges and pizza, and deemed "grotesque," "gross," "diseased." At the same time, Io showed none of the cratering that marred all other known moons.

The reason for Io's garish but smooth countenance was discovered by chance. After *Voyager 1*'s encounter, after the journalistic hordes had vacated the briefing rooms and most of the exhilarated but exhausted mission scientists had either dragged themselves home or taken a much-needed morning break, Linda Morabito, a twenty-six-year-old mathematician, remained at her console. Her task was to find a particular guide star that mission engineers would use to set *Voyager 1*'s course for Saturn. As she fiddled with a routine photograph of Io taken by the probe the day before, an umbrella-shaped crescent blossomed into view above the moon's rim. At first she thought another of Jupiter's satellites might be aligned behind Io, but she quickly discounted that possibility. The crescent looked very like a cloud.

Morabito enlisted the aid of several colleagues still at the lab. By afternoon they were certain that the umbrella was not a smudge on the lens or a glitch in the computer. It had to be a cloud. Next Morabito collared a couple of stray researchers from the imaging team, who agreed, yes, it was a cloud, probably of volcanic origin, rising above a volcanic landform.

Brad Smith, who got the news of extraterrestrial volcanoes almost immediately by telephone, later called Io one of the most bizarre objects in the Solar System. In the days following Morabito's find, team scientists pored over a host of images and spotted seven more eruptions spouting plumes of material high into the moon's thin atmosphere. The largest plume rose about 175 miles above a gigantic hoofprint-shaped feature 420 to 600 miles across. The team spotted great, fan-shaped lava flows as well as dark tentacles extending from flat, volcanic calderas. Geologists had seen nothing like it anywhere else.

This was not to say that no one had imagined it, however. Stanton Peale, a physicist at the University of California, Santa Barbara, and Patrick Cassen

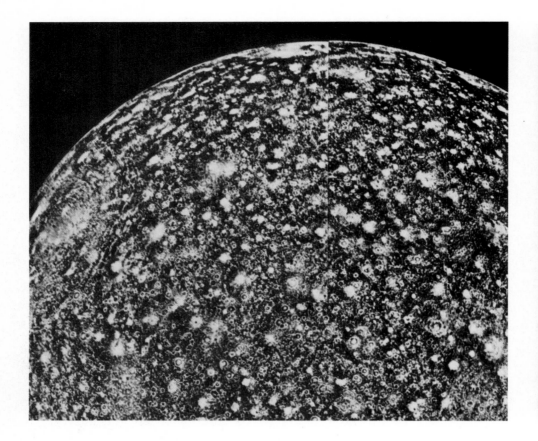
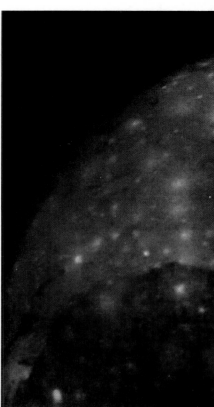

and Ray Reynolds, both at NASA's Ames Research Center, had predicted volcanism on Io in a timely article published in the journal *Science* three days before the *Voyager 1* flyby. According to their calculations, the gravitational tugging of Jupiter, combined with that of Europa and Ganymede, would produce tensions of such magnitude in Io that the little moon would heat up. The energy generated by this tidal flexing, they figured, would be equivalent to 2,400 tons of dynamite exploding every second. Io's core would have to be molten, overlaid by a thin crust, and marked by "events of sufficient energy to penetrate" this outer shell—in other words, volcanic eruptions.

Ejecting material at extremely high velocities, Io's volcanoes *(pages 61-64)* spew out some 10,000 tons of sulfur and sulfur dioxide a second. Altogether, the vents may pump out 100 billion tons of matter each year. Most of this falls to the ground in a blanket of particles—a perpetual refurbishing that gives Io its young, crater-free complexion. All this activity also helps explain how interactions between Io and Jupiter's magnetic field produce radio emissions. The ejecta that escapes the little moon's gravity is strewn around its orbital path in a doughnut-shaped trail called a torus *(page 68)*. As Jupiter's magnetic field sweeps past, the particles in the torus induce disturbances that register as static on earthbound radio telescopes.

NO TWO ALIKE

Io was *Voyager 1*'s first Jovian moon, caught on March 5 during the probe's closest approach. The craft's cameras then turned to the other large satellites first discerned by Galileo in the seventeenth century—Ganymede, Callisto, and Europa. When *Voyager 2* arrived, it trained its slow-scanning video

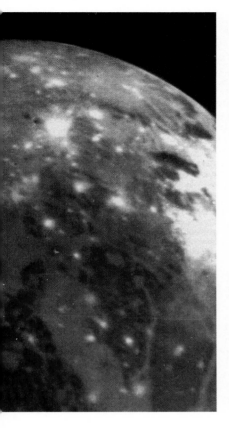

Among the images sent back from *Voyagers 1* and *2* were some that revealed the idiosyncrasies of Jupiter's large moons. For instance, Callisto *(above, left)* is the most heavily cratered object in the Solar System, indicating great age. Its most prominent feature is the Valhalla Basin, which can be seen at the left edge of the photo.

Largest of the moons, Ganymede *(center)* is covered with a crazy quilt of craters, basins, and grooves; its crust may have shifted about considerably in its youth. The darkest regions have weathered the most and are considered to be the oldest.

Unlike the other big moons, diminutive Europa *(right)* is smooth and featureless. The fine lines hatching its surface may be cracks in a crust of ice, filled in with darker material from below.

cameras on those moons, as well as on Amalthea. Between them, the two probes revealed a family of worlds as different from one another as Io appeared to be from everything else.

The ancient face of Callisto is a record of eternal battering dating back several billion years, when the Sun was young. Thousands upon thousands of impact craters, one atop the other, make this moon the most densely cratered object ever seen. Callisto also bears the mark of a massive impact. Called the Valhalla Basin, it is a huge crater about 1,800 miles in diameter, containing a series of concentric rings. Planetary geologists speculate that a mantle of slushy ice or water underlies Callisto's ravaged crust and that the circles in Valhalla Basin are the frozen record of a long-ago impact, rippling outward from the bull's-eye. These old scars have softened over the ages as the malleable crust has sagged and sought its own level. Beyond the cratering, the moon is quite even in its terrain. There are no cracks, mountains, or extinct volcanoes. This absence of any sign of geologic life leads scientists to conclude that Callisto has been cold from birth, a crust of rock and ice covering a dead interior, enduring an eternity of pounding by meteorites and a barrage of orbiting debris left over from the formation of the Jovian system.

If Callisto seems to have no geologic history, Ganymede, larger than Mercury, presents a complex one. Encased in a cracked crust of ice, Ganymede is believed to be about half water by weight, with perhaps a small core of silicate rock. Pale terrain covering about half of the moon's surface seems to have been combed by a giant rake. Fully a third of the far side is marked by a dusky crust known as Galileo Regio, the only feature on any of these satellites large enough to be seen by telescopes on Earth. (Like most moons, Ganymede

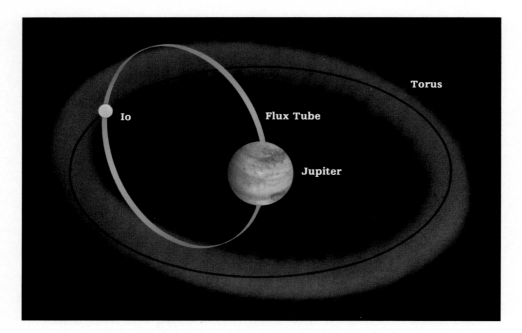

Jupiter's satellite Io orbits the planet within a doughnut-shaped torus of hot, electrified gas, or plasma, consisting mostly of sulfur and oxygen ejected from Io at the rate of ten tons per second. The interaction of the two bodies also produces an immensely powerful river of electricity, known as a flux tube, along Jupiter's magnetic field lines. At five million amps, the current contains about seventy times the power produced by all the generators on Earth.

always shows the same face to its planet.) Parallel mountain chains, rifted by faults, run for hundreds of miles. Planetary geologists think the moon may once have seen tectonic activity, the planet-shaping movement of rigid crustal plates. The rifts shearing its mountain ranges probably cracked open when the surface expanded to vent heat escaping from Ganymede's ancient core.

Europa is stranger still. This smallest Galilean moon has neither mountains nor valleys, craters nor volcanic calderas, but is as featureless as a bowling ball, crazed like a porcelain vase—or a cracked eggshell, as one waggish observer demonstrated with a hard-boiled egg in the JPL press room. Why Europa is the smoothest object known in the Solar System is unclear. According to one controversial scenario, it underwent a period of intense heating at some point in its evolution. Water bubbled from its interior and froze into a thick sheath. As the sheath expanded because of further heat release, water welled up into the cracks and refroze, creating the observed mesh of dark lines. Some biologists have proposed that life forms may swim in the murky ocean that probably underlies this moon's icy coating.

For all the revelations of the two *Voyagers,* Jupiter and its moons still present an abundance of mysteries. But some of them may soon be solved. In the 1990s, a new NASA probe is scheduled to fly to the Jovian system for a closer look. *Galileo,* as the mission is called, will approach Jupiter as a single spacecraft. Six months before its six-year voyage ends at the planet, the main body of the craft will eject an instrumented probe on a long, ballistic fall into a world that has thus far been hidden to robot sensors. The main ship will listen to the probe and relay its findings earthward, until the sensor package is crushed into silence by the great pressures within the Jovian atmosphere.

Then *Galileo* will power itself into the first of a series of orbits, monitoring Jupiter and its moons in the same way weather satellites watch Earth, using imaging devices capable of seeing objects whose sizes are reckoned in mere yards, not miles. For two years, and possibly longer, the giant of the Solar System will become a cosmic laboratory, a tourable museum of primordial materials and events, spinning below the eyes of the robot from Earth.

ANATOMY OF A GIANT

ore massive than all of the other planets put together, Jupiter is easy to see from the Earth with the naked eye. But in their efforts to understand this king of the Solar System, astronomers have always been forced to rely on indirect means. Like detectives poring over chips of paint and shards of glass, they solve pieces of the puzzle by inference and deduction. As long ago as the late 1600s, Isaac Newton used observations of the orbits of some of Jupiter's sixteen moons to estimate the planet's mass, a figure now known to be more than 300 times that of Earth's. Similar studies yielded measures of the giant's diameter (more than ten times Earth's). With these calculations in hand, scientists gauged Jupiter's average density as being only one-fourth that of rocky Terra.

Moving from clue to clue, astronomers arrived at a picture of a planet that is, like the Sun, made up mostly of hydrogen, with a little helium and a pinch of assorted other elements, among them carbon and nitrogen. Indeed, if Jupiter had been only eighty times larger—a pittance on the cosmic scale—temperatures at its core would have caused it to ignite like the Sun itself.

Although inference continues to play a major role in ferreting out Jupiter's secrets, space probes such as the *Pioneer*s and *Voyager*s have produced a treasure trove of new clues. As shown on the following pages, the Jovian world is one of extremes, from its pressure-cooker core and years-long storms to the magnetic fields that permeate space for millions of miles around it.

One and a half times Earth's diameter, yet ten to thirty times Earth's mass, Jupiter's core seethes at temperatures around 55,000 degrees Fahrenheit, the consequence of a pressure of about a hundred million atmospheres.

Core

Unknown on Earth, liquid metallic hydrogen forms under extreme pressures at Jupiter's depths. Hydrogen molecules *(below, left)* are packed so tightly that the attraction between them overpowers the forces binding each together, and the molecules and their atoms break up. Free protons form an irregular grid *(center)*. The flow of electrons through the grid creates the electrical conductivity of a metal and generates a magnetic field.

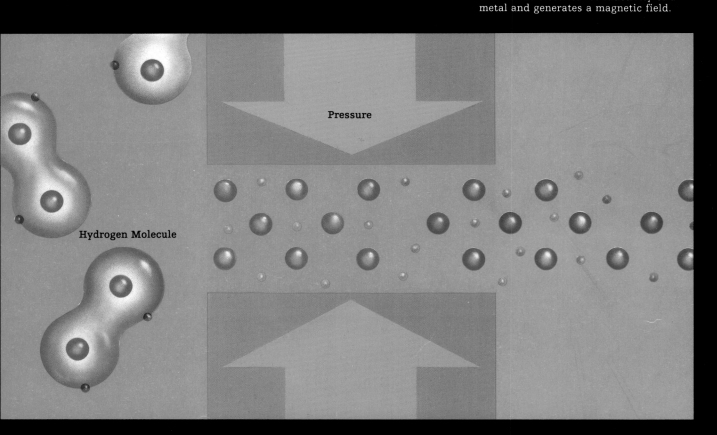

Pressure

Hydrogen Molecule

Surrounding the core is a layer of liquid metallic hydrogen about 25,000 miles thick. At temperatures from 20,000 to 50,000 degrees and pressures in excess of three million atmospheres, hydrogen becomes electrically conductive.

Next is a layer about 13,000 miles thick, made of molecular hydrogen and helium. As pressure falls into the range of tens of atmospheres and temperature drops to just over 100 degrees, these components change from liquid to gas.

A cloud cover less than 100 miles thick swathes Jupiter in patterned shades of dull orange and brown. At the cloud tops, temperature and pressure may decrease to minus 250 degrees and two-tenths of an atmosphere.

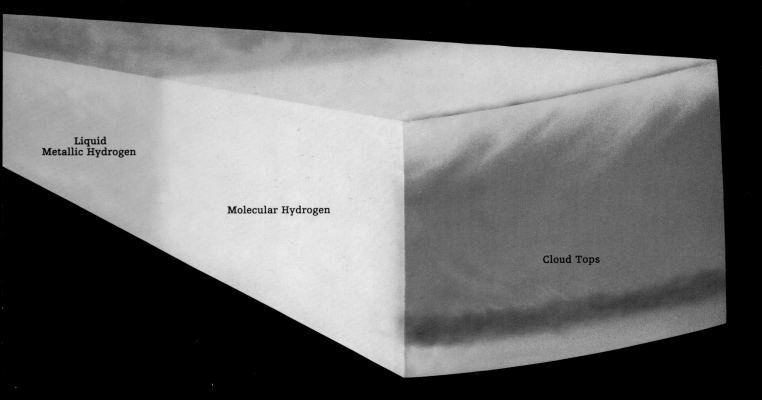

Liquid
Metallic Hydrogen

Molecular Hydrogen

Cloud Tops

From the Inside Out

Like Saturn, Uranus, and Neptune, Jupiter is a highly fluid planet. Heat and pressure within the giant globe dictate where specific chemical processes occur, splitting the interior into distinct layers and generating powerful, often violent currents *(pages 74-75)*.

At the planet's core, which is situated some 44,000 miles below its topmost clouds, the pressure is as much as a hundred million times the air pressure at sea level on Earth (a measure known as one atmosphere, equal to 14.7 pounds per square inch). Only half again the diameter of the Earth, the core accounts

for a mere four percent of Jupiter's mass. Scientists speculate that the heart of the planet is composed chiefly of iron and silicates, but they do not know whether it is solid or liquid.

Around this dense center lies a thick layer of liquid metallic hydrogen, an exotic form of the element that is electrically conductive *(left)*. Hydrogen and helium in more conventional incarnations have their place farther out, gradually merging into an atmosphere of molecular hydrogen mixed with various compounds to produce Jupiter's distinctive cloud display.

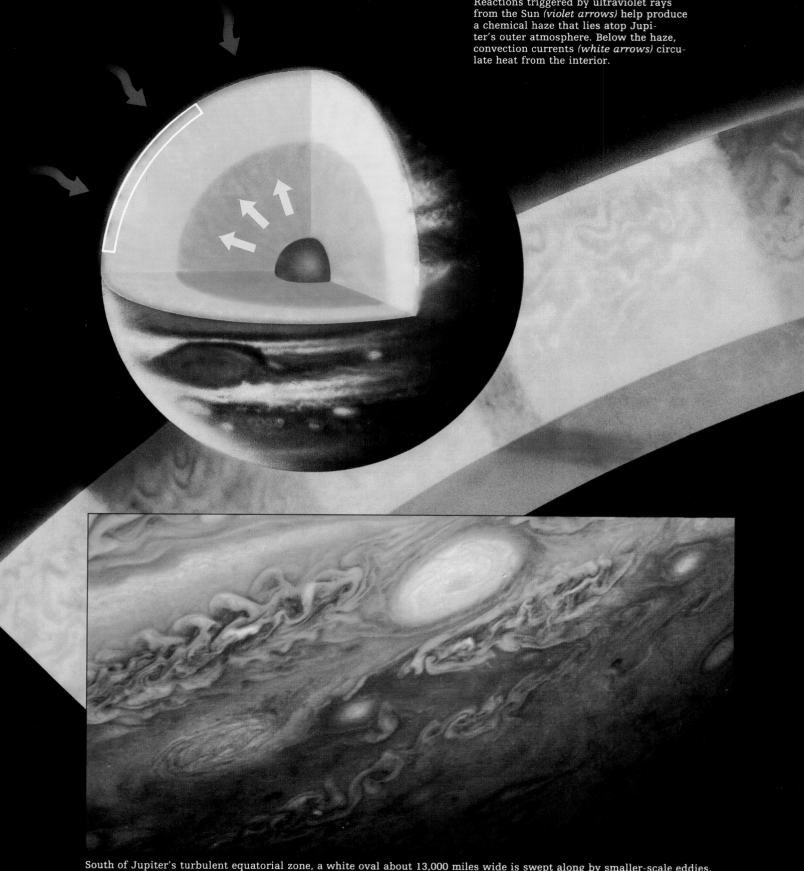

Reactions triggered by ultraviolet rays from the Sun *(violet arrows)* help produce a chemical haze that lies atop Jupiter's outer atmosphere. Below the haze, convection currents *(white arrows)* circulate heat from the interior.

South of Jupiter's turbulent equatorial zone, a white oval about 13,000 miles wide is swept along by smaller-scale eddies.

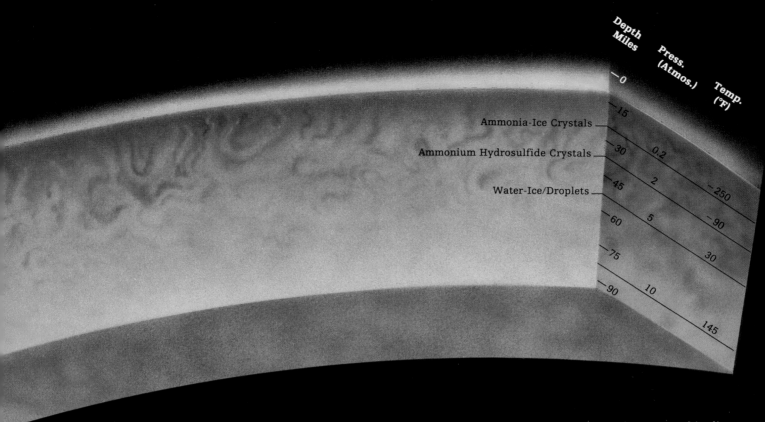

Depth Miles	Press. (Atmos.)	Temp. (°F)
0		
15		
Ammonia-Ice Crystals — 30	0.2	
Ammonium Hydrosulfide Crystals — 45	2	250
Water-Ice/Droplets — 60	5	90
75		30
90	10	
		145

Scientists believe that temperature and pressure gradients have produced a three-part stratification in the chemical composition of Jupiter's outer atmosphere. The innermost level, beginning about fifty-five miles below the cloud tops, may be made up of water droplets and ice crystals at pressures of roughly five atmospheres and temperatures of some 30 degrees Fahrenheit. About twenty miles above the water and ice lies a layer of ammonium hydrosulfide crystals where temperatures hover at about minus 90 degrees. At the outermost level, estimated to be fourteen miles thick, temperatures plunge to minus 250 degrees and pressures to about two-tenths of an atmosphere. In this environment ammonia crystals condense out of the chemical soup.

OF CHEMISTRY AND CLOUDS

Like the thin skin of an apple, a layer of atmosphere less than a hundred miles thick lies between the frigid reaches of space and the vast bulk of hot, pressurized gases that make up Jupiter's interior. In the late 1970s, *Voyagers 1* and *2* examined this outer skin in detail by spectroscopy, revealing an atmospheric witches' brew of molecular hydrogen and helium laced with ammonia, methane, ethane, and traces of other compounds. Arrayed in three levels *(above)*, this chemical caldron wraps the planet in bands of dull brown and orange.

Although scientists have made progress in understanding the composition of Jupiter's outer atmosphere, they remain uncertain about some aspects of its dynamics. The interplay of strong horizontal winds and vertical heat currents *(pages 74-75)* can account for the planet's familiar canvas of swirling eddies and roiling storms *(opposite)*, but the source of its multi-hued palette is a matter of debate. Astronomers once thought the colors were stratified, with lower-level brown and blue clouds peeking through holes in higher white and red clouds. The planet's distinctive coloration is now believed to originate in the uppermost ammonia cloud layer from chemical agents not yet fully identified.

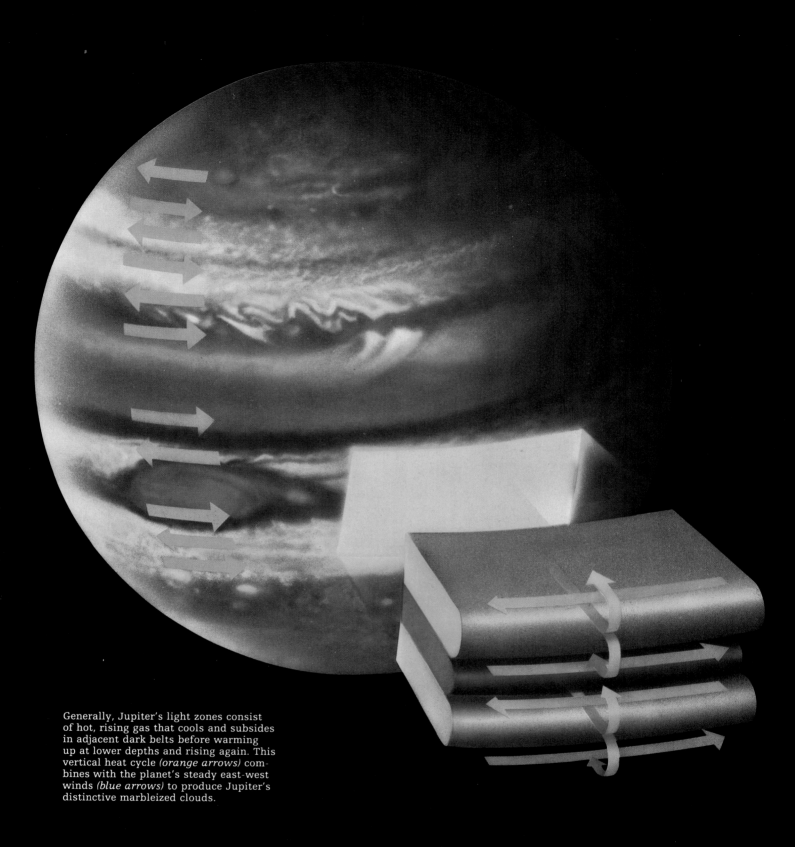

Generally, Jupiter's light zones consist of hot, rising gas that cools and subsides in adjacent dark belts before warming up at lower depths and rising again. This vertical heat cycle *(orange arrows)* combines with the planet's steady east-west winds *(blue arrows)* to produce Jupiter's distinctive marbleized clouds.

spinning on its axis once every ten hours, Jupiter stretches its atmosphere into east-west flows *(opposite)* that combine with convection currents to produce light and dark bands, each thousands of miles wide. The bands may shift and swirl over days, weeks, or years. Continuous wind currents *(blue arrows)* blow at 200 to 330 miles per hour.

A WINDBLOWN WORLD

Three centuries of telescopic observations and recent space probe studies suggest that Jupiter's weather, like Earth's, is a series of short-term variations on long-term patterns. Not surprisingly, however, Jupiter's weather happens on a much larger scale. As on Earth, the Coriolis effect—the tendency of masses in the atmosphere to drift sideways as a planet rotates—results in permanent east-west winds. But Earth has only one jet stream in its northern and southern hemispheres, along with a matching pair of lower-altitude trade winds, whereas Jupiter is swept by about a dozen prevailing winds and by atmospheric eddies large enough to swallow at least two Earths *(right)*.

Scientists theorize that Jupiter's rapid ten-hour rotation period and its lack of a wind-resisting solid surface may contribute to the blustery weather. However, they remain mystified as to how the winds spring up in the first place. On Earth, winds are generated and maintained by large temperature differences—more than 100 degrees Fahrenheit—between the poles and the equator, but on Jupiter the difference is roughly 5 degrees. One possibility is that the Jovian winds are fed indirectly by vertical convection currents driven by the planet's internal heat. With a motion that has no earthly counterpart, the currents could power the atmosphere's swirling eddies. Then, as the eddies break up—after days or decades, depending on their scale—they could transfer kinetic energy to the planet-girdling airflows.

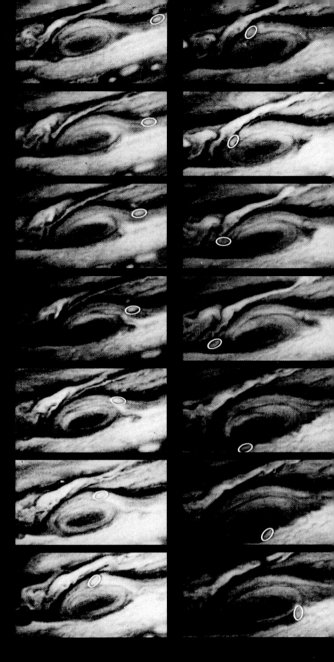

In this series of images taken by *Voyager 1*, a Jupiter-scale anticyclone known as the Great Red Spot seizes a small white oval approaching from the northeast *(top left)*. Over the course of about six Jovian days *(left column, then right)*, the spot whirls the smaller feature counterclockwise and then ejects it—somewhat reduced in size—to the southeast *(bottom right)*. Flanked by a west wind and an east wind and rotating at more than 200 miles per hour, the spot is at least 100—and perhaps as much as 300—years old.

An Invisible Cloak

The most dramatic by-product of Jupiter's inner sea of electrically conductive liquid metallic hydrogen is a magnetic field thousands of times stronger than Earth's. It envelops the planet in a so-called magnetosphere, a region where the magnetic forces hold sway over all electrically charged particles. Shown here in cutaway, the magnetosphere is compressed on one side by the solar wind, a plasma of particles that streams from the Sun at 250 to 500 miles per second. On the opposite side of the planet, it stretches out into a magnetotail that is believed to reach almost to Saturn's orbit. If the magnetosphere were visible from Earth, it would dwarf the full Moon in the night sky.

Even the solar wind is daunted by this powerful Jovian shield. At the magnetopause, the limit of Jupiter's magnetic dominance, the solar plasma parts like water at the bow of a moving ship *(right)*. Particles that flow through this so-called bow-shock region brake abruptly, unleashing the energy of their motion and rendering the plasma twenty times hotter than the Sun in a zone called the magnetosheath.

Although most of the solar wind surges around the magnetosphere, small quantities of plasma make it across the magnetopause to be trapped by belts of radiation. In one instance, the solar particles join with matter that has been spewed into space by volcanoes on the Jovian moon Io to form a doughnut-shaped region *(page 68)* mostly consisting of sulfur and oxygen ions. This region in turn leaks particles that spread out into a thin, electrically conductive plasma sheet *(pink)*.

Magnetosheath

Magnetopause

Solar Wind

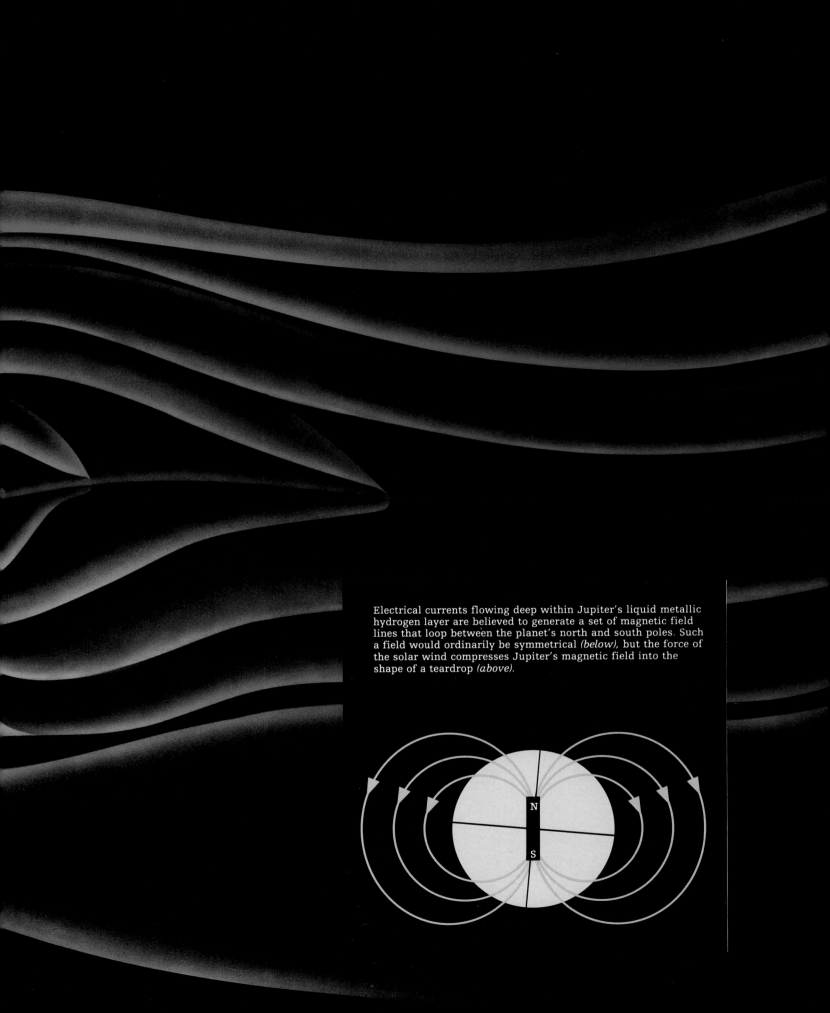

Electrical currents flowing deep within Jupiter's liquid metallic hydrogen layer are believed to generate a set of magnetic field lines that loop between the planet's north and south poles. Such a field would ordinarily be symmetrical *(below)*, but the force of the solar wind compresses Jupiter's magnetic field into the shape of a teardrop *(above)*.

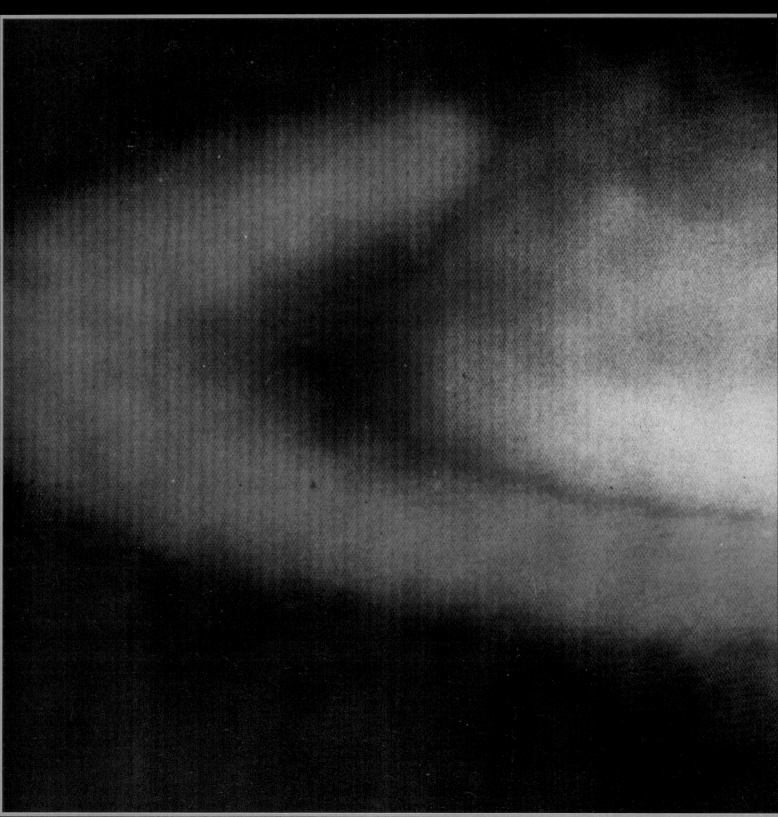

A composite portrait of Saturn made from images taken at two different wavelengths of infrared reveals the warmth radiating from the planet's core and the iciness of its trademark rings.

ore than four years after *Pioneer 11* used Jupiter's gravity like a slingshot to lob itself a billion and a half miles across the Solar System, the space probe began its accelerating fall toward Saturn. As it approached the pale, sand-colored giant in 1978, its earthbound masters at NASA's Ames Research Center faced a tough decision: What route past the planet should *Pioneer 11* follow to return the most useful information? The majority of the instruments aboard the craft were designed by physicists to gather data on the subatomic particles and electromagnetic fields present in a planet's environment. To get close to Saturn, however, the probe would have to storm the diaphanous but potentially deadly Maginot Line of concentric rings orbiting in the plane of the planet's equator.

Astronomers had designated the various rings by letters of the alphabet, assigning the letters in the order of the rings' discovery. The outer edge of the 9,300-mile-wide A ring lay about 44,000 miles above the planet's atmosphere. Circling inside the A ring—and separated from it by a 3,000-mile-wide gap called the Cassini Division, so named for the seventeenth-century astronomer who originally spotted it *(page 87)*—was the B ring, about 16,000 miles wide. Next to be sighted was the C ring, about 12,000 miles wide and thought to be the innermost of the set.

In theory, the best path for *Pioneer 11* would be through the 8,000-mile gap between the C ring's inner edge and the top of the atmosphere. But even that was risky. Some scientists had reported an additional ring, tentatively identified as D, between Saturn and its C ring. Sending the probe through that narrow opening could spell disaster. A collision with a ring particle, even one less than a millimeter in diameter, could seriously damage the speeding spacecraft. The alternative, to send *Pioneer 11* outside the orbit of the A ring, carried risks as well—namely, the possibility of still another ring, designated E, beyond the A ring.

The jury-size team of scientists and engineers supervising the Pioneer mission voted eleven to one for the inside track, dangerous as it was, in order to get the best possible data. But the recommendation did not sit well with Tom Young, the Virginian who was then NASA's director of planetary programs and had the final word on such decisions. He saw clearly that *Pioneer*

11 was needed as a scout, both for the planned Saturn encounters of the two Voyager spacecraft in 1980 and 1981 and because of a related decision just made by NASA. *Voyager 2* was to continue on to Uranus and Neptune, a journey it could not manage without drawing energy from Saturn's gravity. Aiming *Pioneer 11* at a point outside the rings would send the probe on a trajectory close to the one *Voyager 2* would have to follow in order to tack toward Uranus. *Pioneer 11,* Young decided, would have to find out if that path was open. For the sake of its successor, the aging probe would be denied its best shot at lifting Saturn's inscrutable mask.

PLANETARY REVIVAL

The mask had been held firmly in place from the time humans first noticed the slow-moving beacon that seemed to mark the rim of the universe. Although Saturn, with a diameter of about 75,000 miles, is the second largest planet in the Solar System, it orbits the Sun at a mean distance of about a billion miles, too far away to be seen very clearly from the vantage point of Earth. Light reflected and scattered from the rings makes the planet itself difficult to view even through large, modern telescopes. Saturn's twenty-nine-and-a-half-year orbit presents further problems; tracking a single revolution of the sixth planet could consume the entire working career of an astronomer on the third.

Even figuring out the length of a Saturnian day was no easy matter. With Jupiter, astronomers could time the periodic appearance of such distinctive features as the Great Red Spot. Saturn was less cooperative. Although very dim latitudinal bands, or belts, were seen as early as 1676, the first observation of anything on Saturn resembling the Jovian spots did not come until the 1790s, when William Herschel tracked luminous features in the rings and markings on the planet itself to calculate Saturn's rotation at ten hours, sixteen minutes. Because Herschel made his measurements at the limit of what his telescope could resolve, his findings were regarded as uncertain. But in 1876, an astronomer at the U.S. Naval Observatory examined Saturn with the greatly improved equipment of his day and reported that Herschel's estimated rotational period had been short of the correct figure by only a quarter of an hour.

Saturn's almost featureless appearance stimulated speculation about what might lie beneath the dense cloud cover. Theories ranged from a world of pure gas to a surface flowing with molten lava. Rupert Wildt's identification in 1931 of methane and ammonia absorption lines in the spectra of the far planets dismantled the wilder scenarios. Instead, Saturn simply seemed to be a smaller version of Jupiter, consisting mostly of hydrogen. Applying Wildt's models for Jupiter's structure, some astronomers hypothesized that Saturn harbored a relatively compact core of rocky silicates surrounded by layers of water ice and a dense atmosphere consisting of hydrogen, helium, methane, and ammonia.

In a sense, these insights marked the stirrings of new life in planetary astronomy, which in the first half of the twentieth century had been made to seem somewhat pedestrian against a more exotic study of stars, galaxies, and the large-scale structure of the universe. The study of planets had also suffered from an earlier generation's love affair with Martian canals and lunar civilizations. Hence, during this period, only a very few professional astronomers continued to focus on the worlds of the Solar System. Among them was Gerard Peter Kuiper, a Dutch-born scientist who would, in effect, become the father of modern planetary astronomy.

INVENTING INFRARED

Trained at the University of Leiden, the Netherlands' center of excellence in astronomy, Kuiper emigrated to the United States after winning his doctorate in 1933. He first took an appointment at Lick Observatory in California, then three years later joined the staff at the Yerkes Observatory in southern Wisconsin. There, and at the affiliated McDonald Observatory in Texas, he set about virtually inventing the discipline that would be known as infrared astronomy. In the process, he led the study of planets away from its fantasizing habits and back to real science.

The infrared spectrum runs from wavelengths of about one micron (one millionth of a meter)—just beyond what the human eye can see—to the shorter wavelengths of the radio spectrum. Energy detectable in the so-called near infrared, from about two to forty microns, corresponds to temperatures ranging over a few hundred degrees on either side of the freezing point of water. Interstellar matter, clouds of galactic dust, and planets all radiate strongly in the infrared, but because almost everything in the universe emits some heat, distinguishing the sources of infrared radiation is difficult. Kuiper had his work cut out for him.

In 1944 he applied his infrared techniques to Titan, the largest moon of Saturn. Using the 2.1-meter reflector at McDonald Observatory, he obtained spectra that showed the distinctive absorption band of methane, confirming that, in a universe where moons were supposedly dead spheres of ice and rock, Titan had an atmosphere. The presence of an atmosphere suggested that the surface temperatures of Titan could be considerably warmer than those of airless bodies far from the Sun: Like Earth's atmosphere, the gases enveloping Titan would trap heat. And if the moon had an atmosphere and was warm, it might also support some form of life. Decades later, when planning the trajectories for the two Voyager probes, NASA scientists would have a choice of sending *Voyager 1* close to Titan or on to an encounter with Pluto. They chose Titan.

A titan himself, Kuiper left enormous footprints across the early history of planetary studies. He discovered a Uranian moon, Miranda, in 1948 and a Neptunian moon, Nereid, the following year. More than two decades later he was still making important contributions. In 1970 his spectroscopic work confirmed what astronomers had suspected: Saturn's rings were mostly ice and ice-encrusted rock. Kuiper's greatest legacy, however, may be the generation of men and women drawn to the science he had revitalized, a generation that formed the youthful heart of teams like Pioneer's. Kuiper lived to see them launch their probes. When he died in December 1973, *Pioneer 11* was outbound on the path that would ultimately lead to Saturn.

DESCENT INTO DANGER

As the little craft plunged to within a million miles of the planet on August 30, 1979, mission controllers at Ames Research Center watched for the sudden shift in readings—changes in magnetic-field intensity and the numbers and speeds of charged particles—that would mark the probe's crossing of Saturn's bow-shock region. Later than expected, the sensors detected such variations, then lost them, as the planet's magnetosphere billowed and collapsed in the gusts of particles streaming from the Sun. Not until about 5:00 p.m., when the probe was only about 635,000 miles from Saturn, did *Pioneer 11* make its final crossing of the pulsating boundary. A few short hours ahead lay the far more dangerous crossing of the ring plane.

As decided by Tom Young, that perilous passage would be made at a distance of 71,000 miles from Saturn's cloud tops. With the probe traveling at 53,000 miles per hour and gaining speed, the event would last only a fraction of a second, but for the mission team at Ames, this was a compact eternity. The crossing was predicted for 9:02 a.m. on September 1, 1979. The signal—or the silence—that followed would take another eighty-six minutes to reach Earth. If *Pioneer 11* was destroyed, its signal would cease at 10:28. As the probe fell toward a point about 26,000 miles outside the A ring—a point that might or might not be occupied by ring material—the men and women at Ames, a billion miles away from their plunging surrogate, could only wait in their hushed control room and stare at their clocks, watching 10:28 approach.

The crucial moment came and went, and *Pioneer 11*'s signals continued to stream into the network of deep-space radio antennas arrayed in the Mojave Desert. But eight more minutes crawled by before project manager Charles Hall, known for his technical prudence, allowed himself to be convinced that the probe had really survived its dangerous passage. Finally, he announced

Luminous as a candled egg, brown-ringed Saturn fills this August 1979 image transmitted by *Pioneer 11*, the unmanned craft that gave humanity its first clear view of the sixth planet. The probe's cameras and instruments discovered one small moon, the narrow F ring, and a planetary magnetic field second only to Jupiter's in size.

to his breathless team over the public address system, "I think we made it."

More than *Pioneer 11* had survived. The successful crossing also meant that the hope of exploring the more distant outer planets had not died in the rings of Saturn. Now, amid calls of "On to Uranus!" scientists in California settled down to the business of probing a new world.

When *Pioneer 11* shot through the plane of Saturn's rings, accelerating toward 70,000 miles per hour for its close encounter, the probe's imaging system was, in effect, blinded by the spacecraft's speed. But James Van Allen, the physicist whose satellite experiments twenty years earlier detected the radiation belts that bear his name, had engineered an alternative way for *Pioneer 11* to see. The technique, dubbed particle-beam astronomy by Van Allen, employed the same occultation effect that had led *Pioneer 11* to sense the presence of Jupiter's rings. As the spacecraft hurtled toward the planet, the rings blocked the stream of subatomic particles that had otherwise been reaching its sensors; changes in the counts of energized particles within the shadow region revealed the presence of rings, gaps, and anything else near the probe.

By this ingenious method, *Pioneer 11* tentatively verified the presence of the elusive E ring, which stretches for nearly 180,000 miles beyond the well-defined A ring but whose dust particles are so widely scattered that the probe was able to speed through them unscathed. *Pioneer 11* also detected two new rings—a narrow strand, designated F, just beyond the outer edge of the A ring, and a broad band, designated G, inside the E ring. Dust-particle readings also suggested that even the apparent gaps in the rings were not entirely devoid of material.

For the most part, Van Allen's trick yielded the kind of results mission controllers expected. But one reading was downright scary. Moments after *Pioneer 11* crossed the ring plane, the count of subatomic particles abruptly dropped to levels comparable to the faint background radiation on Earth, the lowest count, Van Allen later mused, "since we were sitting on the launch pad at Cape Canaveral." Something had cut between *Pioneer 11* and the streaming particles, producing a kind of guillotine effect.

"We had to be close to something pretty large," commented project scientist John Wolfe, a bearded Ohioan known as "Dr. Sardonicus" to reporters. "It's very intriguing." Later, mission scientists determined that *Pioneer 11* had come within a few thousand miles—a hairline, in space terms—of rediscovering 1979S1, a moon first seen the day before, the hard way.

Skimming only a few thousand miles below the ring plane, *Pioneer 11* tracked around the huge planet, gathering momentum for its escape into the emptiness beyond. As its path took it behind Saturn, the occultation of the craft's radio signal allowed mission scientists to study the planet's atmosphere. Then, four hours after flashing through the ring plane on one side of Saturn, *Pioneer 11* rose through the plane at the same distance on the opposite side, once again without harm. Tom Young was jubilant. "We can report to Voyager, 'Come on through, the rings are clear.' "

Distinct shadows of the moon Tethys and the A, B, and C rings mark the pale, nearly featureless surface of Saturn in this November 1980 scene captured by *Voyager 1.* Tethys itself hovers beside the planet; Dione, another icy inner satellite, orbits at far right.

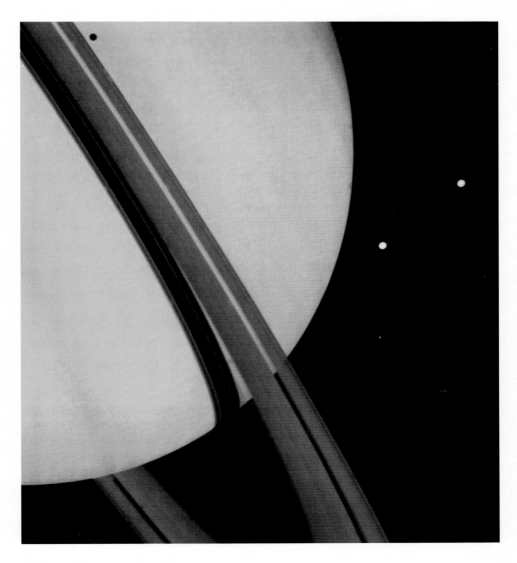

PORTRAITS BY PROBE

Long before *Voyager 1* "came on through" fourteen months later, preliminary data sent back to the Pasadena-based Jet Propulsion Laboratory by the spacecraft made it clear that Saturn was about to be seen in dramatically greater detail than ever before. In October 1980, when the probe was still more than 30 million miles from the planet, *Voyager 1*'s high-resolution cameras reported a host of new and complex phenomena within the rings. Each incoming image shattered conventional belief. *Voyager 1* confirmed *Pioneer 11*'s hint that there were distinct bands of material within some of the classical gaps and went on to find new gaps where none had been observed or predicted. But to JPL's Rich Terrile, working his first deep-space mission, the most puzzling of the images showed long, dark, fingerlike projections radiating outward across the rings, like the spokes of a wheel. Because the chunks of ring material all orbit at their own speeds, such a structure should not, theoretically, be able to sustain itself at right angles to the rings—and yet, here it was. As they had done almost from the time their birds rose from Florida in 1977, mission controllers made an in-flight change, reprogramming *Voyager 1*'s computers for a longer look at these impossible apparitions.

They also focused on the F ring, another defier of conventional belief. According to theory, the ring material should have dispersed into space. Instead, the ring's edges were sharply etched against the darkness. What kept this flock of particles in such a crisp formation? The apparent answer came in late October, when imaging-team scientists analyzing ring photos spied two small moons, one orbiting just beyond the outer edge of the F ring, the other hugging the ring's inner edge. Initially labeled S-13 and S-14 as the thirteenth and fourteenth moons of Saturn (later they would be named Pandora and Prometheus), they quickly became known as the "shepherding satellites." It seemed likely that the two small moons gravitationally herded the F ring particles into tightly confined orbits, thus controlling the ring's size and shape *(pages 106-107)*.

The closer *Voyager 1* got to Saturn, the more rings it saw. By November 7, imaging-team leader Bradford Smith had counted 95 separate rings. A week later, Rich Terrile had increased the count to more than 500. And some of these newly discovered rings were more than a little strange. Within the Cassini Division, the well-defined gap between the A and B rings, and also within a small gap in the C ring, *Voyager 1* detected narrow rings that clearly were not circular. "The mystery of the ring structure gets deeper and deeper," commented Smith. "The thing I perhaps least expected to see was eccentric rings, and now we have not one, but two."

By November 12, the day of closest approach, Smith could contemplate ring images that were positively amazing. Projected on the screen in JPL's Von Karman Auditorium was a high-resolution shot of the F ring showing that it consisted of two strands of material that seemed to be twisted around one another. Scientists and reporters simply shook their heads, happily bewildered. "Obviously, the rings are doing the right thing," said Smith. "It's just that we don't understand the rules."

As *Voyager 1* dropped safely through the ring plane, skimming to within 77,000 miles of Saturn's cloud tops, the parade of surprises kept coming. Instead of the bland, gauzy atmosphere visible through ground-based telescopes, the space probe surveyed a vista of whorls, scallops, and storms that seemed a muted form of the more dramatic, highly colored features seen at Jupiter. High-speed jet streams arrayed Saturn's clouds in latitudinal belts and zones. Radio experimenters heard the distinctive snap and crack of powerful electrical discharges.

Voyager 1's revelations poured in, saturating the tired but euphoric team in Pasadena. "I'm stunned by the pictures," said Brad Smith, speaking for his colleagues as well as himself. "Perhaps a year from now it will be possible to sort them out, but right now we are simply flooded with new data."

TROUBLE AT SATURN

By the late summer of 1981, however, the mood among planetary scientists had changed drastically. Nine months had passed since *Voyager 1*'s flyby, and *Voyager 2* was fast nearing its encounter with Saturn. But now, as if the

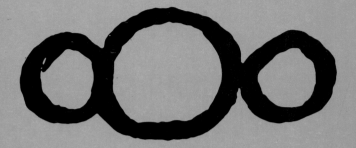

SATURN'S EARS

Upon looking at Saturn through his primitive telescope in 1610, Galileo drew the planet as a big sphere bounded by two smaller ones *(above)*, like a head with large ears. Galileo took the flanking objects to be moons, but when he examined the planet two years later, they had completely vanished. The mystery deepened when the curious appendages later returned to view, and they remained an enigma for more than four decades.

Then, in 1655, Dutch astronomer Christiaan Huygens—having discovered that Saturn possessed an orbiting satellite (the moon Titan)—announced that the planet must also be surrounded by an equatorial ring. So thin and flat was the ring that it disappeared when seen edge on by an earthbound observer, a condition that occurs about every fifteen years in the planet's twenty-nine-year orbit around the Sun.

What Huygens deduced was confirmed in 1675 by the Italian-born astronomer Giovanni Cassini, first director of the new Paris Observatory. He found that Huygens's ring was in fact two rings, separated by a gap now called the Cassini Division. Furthermore, he assumed the rings were not solid but made up of a multitude of small bodies revolving independently in concentric orbits.

Over the next two centuries, many of the great scientific figures of the day were drawn to Saturn's mystery. However, except for Scottish mathematician James Clerk Maxwell, whose own investigations in 1857 led him to side with Cassini, most astronomers believed that the rings were rigid disks. Finally, in 1895, James Keeler of Pittsburgh's Allegheny Observatory analyzed Saturn's spectrum and proved that, as Cassini had intuited, Saturn's "ears" were in fact many small objects that were orbiting the planet as obediently as the planets orbit the Sun.

hazards of a ring crossing were not enough, *Voyager 2* was beginning to have financial problems at home.

A new administration's cuts in the federal budget had chopped nearly every new planetary mission, and there was even disquieting talk of plans to "pull the plug" on *Voyager 2*'s scheduled encounter with Uranus in order to save money: The spacecraft might sail past Uranus in 1986, conduct its preprogrammed encounter sequence, and beam the data back to an unlistening Earth. NASA also went ahead with administration plans to ax yet another mission, the Jet Propulsion Laboratory's proposed probe of Halley's comet. Where the era of Pioneer, Viking, and Voyager had been a golden age of planetary exploration, *Voyager 2* was being billed as the last picture show.

Despite the disheartening budgetary realities, *Voyager 2*'s handlers held on to their enthusiasm. The willingness to tinker with their spacecraft at the end of a billion-mile-long communications link still flourished. In the interim since *Voyager 1*'s pass at Saturn, mission controllers at the laboratory had redesigned *Voyager 2*'s encounter program to take a longer look at the first probe's multiple surprises.

The rings remained the primary object of interest, as scientists searched for explanations for their perplexing structure. The favorite theory was that tiny moons, similar to the shepherds flanking the F ring, were hiding within the rings themselves and gravitationally controlling the ring structure. "Moonlets between ringlets cause gaplets," as one reporter put it.

Scientists pored over the ever-better views of the rings, fully expecting to find the embedded satellites. However, even down to a scale of about five miles, no moonlets appeared, a circumstance that forced the astronomers to yet another rethinking of ring theory. Meanwhile, the F ring's mystery deepened. What had looked like interwoven strands to *Voyager 1* appeared to *Voyager 2* as separate, kinked ribbons. Apparently, the so-called braids were an illusion that had been caused by overlaying wavy strands of the F ring. Some scientists attributed the waviness to the passage of Prometheus and Pandora, the two shepherd moons, whose gravitational influence could produce the observed wrinkles.

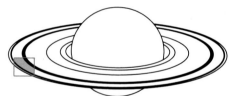

Flanked top and bottom by dense regions in Saturn's A ring, an image from *Voyager 2* focuses on the Encke Division, a 170-mile-wide gap located close to the A ring's outer edge *(diagram)*. The picture, constructed by measuring light from a star passing behind the ring, encodes opaque areas in yellow, less dense zones in red, and relatively clear sections in black. Identified in the 1830s by German astronomer Johann Encke, the gap was found to include several ringlets, shown here as red bands.

The spokes discovered by *Voyager 1* also evaded tidy explanation. A motion picture produced from successive ring images showed the spokes forming, rotating within the rings, and then dissipating within hours. Scientists began speculating about nongravitational forces that might play a role in spoke formation *(pages 104-105)*.

Storm ovals seen in Saturn's atmosphere were another compelling target for *Voyager 2*'s cameras. But to aim them at the ovals, scientists needed to know where the storms were going to be at a particular moment. Caltech's Andrew Ingersoll, a specialist in atmospheric dynamics who had been the weather forecaster for the Jupiter encounter, used images taken at long range to plan the close-encounter sequences, but he admitted, "We have no idea if it's going to work. It's six-week weather forecasting, on Saturn." Ingersoll's apprehension was misplaced, for the predictions turned out to be wonderfully accurate, impressing everyone but the Saturnian weatherman himself. "Saturn," he explained later, "is incredibly cooperative. Each storm stays at its own latitude. Velocities never change." To extrapolate where a storm would be from how fast it was going in one direction was "trivial," Ingersoll said. "We could track the storms from day to day; they always moved in a straight line. We were big heroes for doing what any high school kid could have done."

Meanwhile, real heroics were underway to sharpen *Voyager 2*'s vision during the flyby. The probe was equipped with a photopolarimeter, which could obtain very high resolution images by measuring how light changes as it is reflected or absorbed by an object. But because the identical instrument on *Voyager 1* had failed as the spacecraft raced through Jupiter's sizzling radiation belts, *Voyager 2*'s photopolarimeter had been turned off to protect it from high-voltage surges.

During the early evening of August 25, *Voyager 2* dropped toward its single crossing of Saturn's ring plane, which would be made behind the planet, out of radio contact. As the spacecraft approached the crossing, the photopolarimeter signaled that it was turned on and working and that it had locked onto the light from Delta Scorpii, a star on the opposite side of the rings. One by one, the rings eclipsed the star. The instrument, measuring the intensity of the star's light a hundred times a second, sensed fluctuations caused by the occultation. Later translated into peaks and valleys by computers at JPL, these signals would profile the ring system. From an instrument that had been revived after a four-year nap, the dogged scientists extracted a map so richly detailed that the 16,000-mile-wide B ring could be examined down to the scale of a city block.

Then *Voyager 2* was behind the giant planet, on its own, and—unbeknown to mission controllers in Pasadena—in trouble. When telemetry from the probe resumed, it showed a series of unexpected thruster firings that, according to project manager Esker Davis, "would imply some force hitting the spacecraft." At the same time, Davis reported, the particle and field measurements "just went crazy." Some scientists later concluded that *Voyager 2* had passed through a ring rather than a true opening, one more indication that material probably exists everywhere in the ring plane: No gap is a perfectly safe passageway.

The controllers' worries were far from over. As *Voyager 2* began transmitting images again, the pictures received at JPL showed only empty space. The spacecraft's scan platform, where the all-important cameras were mounted, had begun to stick when the platform turned horizontally, and *Voyager 2*'s computers could not control the cameras. Whether the malfunction was related to the strange events at the ring-plane crossing, no one could say at the time. But with the platform jammed, the cameras could not be pointed without maneuvering the entire spacecraft, an imprecise procedure that could use up some of the attitude-control fuel allocated to the encounter with Uranus. For two harrowing days, mission scientists feared that the little ship had been hopelessly maimed.

But the engineers refused to accept failure. Commanding the spacecraft to move the platform a little, and then a little more, they forced *Voyager 2*'s cameras back toward Saturn for one last shot. It arrived on the monitors at JPL, showing part of Saturn in one side of the frame. Though not much better than a snapshot taken by a small child, it was a crucial image, proving that the long-range massage had at least partially restored the craft to working order. The picture also restored the team's confidence that they could handle emergencies, even though their tether to *Voyager 2* was growing ever more extended and tenuous.

CLUES AND COMPLEXITY

Scientists have only scratched the surface of the mountain of data returned by the Pioneer and Voyager craft, but already they have put together some of the pieces of the complex puzzle that is Saturn. Like Jupiter, which it strongly resembles, Saturn emits more heat than it receives from the Sun—two to three times as much. This extra energy, scientists reason, must be caused by something more than the interior cooling and gravitational contractions of a gas giant, especially since Saturn is both smaller than Jupiter and much farther from the Sun.

Possible clues came from infrared sensings by the space probes. *Pioneer 11*'s infrared detectors confirmed the presence of helium in Saturn's atmosphere and also took the planet's global temperature: At minus 290 degrees Fahrenheit, Saturn was much warmer than it was supposed to be, given its great distance from the solar furnace. Then, further infrared observations from the Voyager probes showed that Saturn's supply of atmospheric helium is depleted compared to the helium contained in Jupiter's atmosphere. Putting these disparate pieces of information together, some scientists speculate that, long ago, Saturn cooled to the point at which helium began to condense out of the upper atmosphere, creating a constant, chilling rain of liquid helium. Like all falling objects, Saturn's helium raindrops store gravitational energy as they drizzle down toward the planet. When they come to a stop, this

Saturn's radio profile, like its infrared portrait *(pages 78-79)*, largely reflects the planet's heat distribution. Strong radio emissions *(red)* at Saturn's center decrease gradually at its edges; beyond the planet's surface, blue tones record weak radiation from the outer atmosphere. Where the cool rings cross Saturn's face, they absorb much of the planet's radiation, forming a yellow line across the image.

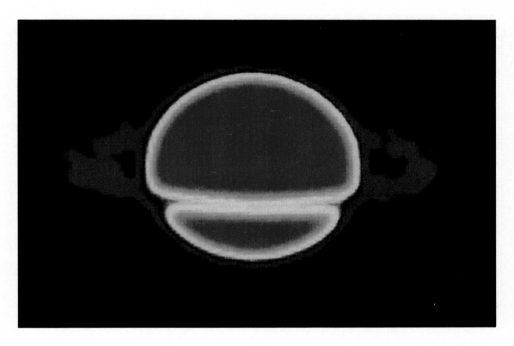

energy of motion is yielded up as heat. (The same is true of ordinary raindrops on Earth.) Accumulating over one or two billion years, the added energy may have replaced the primordial internal heat Saturn lost as it cooled. The warming rains of Saturn have not yet come to Jupiter, where temperatures have only recently dropped to the point at which a helium rain can begin. If Jupiter is just entering a phase already passed by the smaller giant, perhaps Jupiter's future can be read in Saturn's present.

Although the planet's interior remains hidden, scientists have arrived at a consensus theory for its internal structure. Like Jupiter, Saturn is believed to be an enormous ball consisting mostly of hydrogen and helium, compressed by the force of its own gravity. Superficially at least, the two planets are very similar. Just below Saturn's visible clouds, pressure is equal to that at sea level on Earth—one atmosphere, in scientific parlance—and temperatures hover at a frosty 140 degrees Kelvin. As on Jupiter, temperatures and pressures increase with depth. Some 20,000 miles beneath the cloud tops, molecular hydrogen becomes liquid metallic hydrogen at a temperature of 9,000 degrees Kelvin, and pressure reaches three million atmospheres. Saturn is nearly as large as Jupiter—their diameters differ by only about 14,000 miles—but is much less massive, with an average density of just seven-tenths of a gram per cubic centimeter: Saturn would float in water. The planet's core, composed of iron, silicon, and other heavy elements, is thought to be much smaller than Jupiter's and about three times as dense as the average density of Earth, with a temperature of 12,000 degrees Kelvin and a pressure estimated at eight million atmospheres.

The visible part of Saturn, its upper atmosphere, seems to behave like a terrestrial ocean. Responding very slowly to the input of heat from the distant Sun, it apparently draws much of its energy from convective currents welling up from the interior. These currents of hot gases reveal themselves as the storm ovals and whirlpools tracked by the Voyager spacecraft. They apparently energize east-west jet streams, which race along at some 1,100 miles per hour, four times faster than their counterparts on Jupiter.

Winds are not the only form of turbulent weather to be seen at Saturn. The planet also emits electrical bursts with clocklike regularity every ten hours and ten minutes—an indication that their source is rotating along with the ringed globe. Some scientists now believe the discharges originate in a long-lived, 35,000-mile-long lightning storm raging in Saturn's atmosphere about four degrees north of the equator. But the generator behind this enduring tempest remains unknown.

At higher latitudes, a single jet stream dominates the Saturnian subtropics, giving way to alternating east and west flows closer to the poles. The pattern of these winds is virtually identical in the northern and southern hemispheres, leading some scientists to suspect that the symmetry may also extend to great depths. According to one theory, the matching flows in the two hemispheres may simply be opposite sides of concentric cylinders within cylinders of gas, spinning around the planet's axis of rotation.

Space probes detected mostly hydrogen and helium, with numerous trace compounds such as ammonia, methane, phosphine, acetylene, and propane, in Saturn's atmosphere. But none of these explains the muted colors of the planet's imperturbable mask. It may be that a thin layer of haze envelops Saturn, concealing strata that would otherwise appear more like the richly hued, well-defined bands of Jupiter.

MAGNETIC MYSTERY

The other, invisible atmosphere of Saturn's magnetic field has been less forthcoming to theorists. During the *Pioneer 11* encounter, flares on the Sun had strengthened the solar wind, thereby compressing the magnetosphere on its sunward end. But even without a stronger-than-average solar wind, Saturn's magnetic field is surprisingly compact, especially given its strength: While it is only one thirty-fifth the intensity of Jupiter's, the field is 500 times stronger than the Earth's.

The probes also detected a more fundamental magnetic distinction. Unlike any other known planetary magnetic field, Saturn's has an axis aligned almost perfectly with the planet's axis of rotation. Until astronomers made this discovery, theory held that magnetic fields were in part created by the misalignment of magnetic and rotational axes. To have them lined up so neatly sent scientists scurrying for new hypotheses.

Other peculiarities of the Saturnian magnetic field are traceable to the planet's rings and moons, especially Titan. Although Titan's face remained hidden by clouds during *Voyager 1*'s flyby, an occultation experiment employing the spacecraft's radio signal revealed that this moon's atmosphere is even denser than Earth's, but its surface temperature is much colder—minus 290 degrees Fahrenheit—than scientists anticipated. Still more surprising, Titan's atmosphere is composed almost entirely of nitrogen, with relatively small amounts of the methane Kuiper had detected nearly four decades earlier. As Titan orbits through Saturn's magnetosphere, some of the hydrogen and other light elements in its dense atmosphere escape, as they did from

DESCENT TO A PRIMORDIAL OCEAN

Early in the twenty-first century, a conical probe packed with instruments is scheduled to parachute into the murky atmosphere of Titan, largest and most intriguing of Saturn's moons. Scientists can only speculate about the world beneath Titan's opaque cover, but information collected by the 1980 *Voyager 1* flyby indicates that the probe will survey one of the most bizarre landscapes in the Solar System.

If one widely accepted theory is correct, the distant moon is awash in a polluted ocean of hydrocarbons (chiefly liquid ethane and methane) and dotted with islands of water ice, the satellite's crust. A frigid rain of organic compounds created in the Titanian atmosphere blankets the islands with powdery debris and the ocean bottom with a gooey sludge. The faint, faraway Sun gleams dimly through occasional thin patches in the red-brown smog.

This gloomy atmosphere holds great appeal for NASA and European Space Agency scientists planning an unmanned mission to Saturn that is slated for launch in the mid-1990s. For four years, the Cassini mission, as it is named, will explore the chemistry of Saturn and its principal satellite. The reason: Titan's atmosphere may closely resemble that of primordial Earth four billion years ago. Though much colder than the atmosphere of the early Earth, it could yield a treasure trove of information about the predawn of terrestrial life.

Some 300 feet or so from impact, the Titan probe nears the end of its three-hour descent through the Saturnian moon's atmosphere. The five-foot-wide probe will be launched from a spacecraft in orbit around Saturn and carry instruments for studying the chemical composition and structure of Titan's atmosphere as well as the topography and makeup of its icy surface.

Ultraviolet absorption layer. At roughly 250 miles, atmospheric gases absorb ultraviolet radiation.

Optical haze layer. At 190 miles, organic compounds create a thin, hazy layer that scatters and absorbs visible light.

Aerosol layer. At 140 miles above Titan, organic compounds begin to settle out of suspension and stick together, forming tiny particles that float slowly to the moon's surface. The result is a reddish brown smog.

Methane cloud layer. Vaporous clouds of frozen methane may float about 25 miles above the ground, drizzling liquid methane onto Titan's surface.

A Waiting Cradle for Life

The presence of organic compounds—the raw materials of life—in Titan's atmosphere excites scientists more than any other aspect of the distant world. Some researchers even dream of performing large-scale experiments on Titan one day, of heating up acres of its surface to brew a primordial soup of the sort that spawned life on Earth.

Interest in such experiments stems from significant similarities found in scientific models of present-day Titan and primitive Earth. According to these models, the bombardment of atmospheric gases by solar radiation and particles from the body's own magnetosphere might have set off a chain reaction on both worlds, leading to the formation of complex organic molecules (right). The atmospheres of both Titan and primordial Earth would thus contain significant quantities of nitrogen and carbon-based gases. Moreover, neither would produce free, or uncombined, oxygen—that is, oxygen atoms not locked up in carbon monoxide or other simple molecules. Free oxygen would inhibit construction of chemicals such as hydrogen cyanide (HCN), an intermediary in the synthesis of adenine, one of the four nucleic acids that make up the DNA (deoxyribonucleic acid) genetic code.

Once formed, organic molecules in Titan's atmosphere could make chains called polymers, which would clump together into tiny particles that sink slowly to the surface (opposite). If Titan's surface is solid, the accumulated deposits laid down over the last 4.6 billion years may be deep enough to bury a 300-story building—potentially the richest store of organic compounds in the Solar System.

Plentiful as they may be, these compounds probably have not nurtured the beginnings of life because of at least one essential difference between primitive Earth and Titan. Earth basked in heat of more than 100 degrees Fahrenheit, but Titan lies paralyzed at minus 290 degrees, so that water on Titan is locked up as ice, unable to interact with carbon compounds. Some scientists speculate that five billion years from now, when the Sun balloons into a red giant, Titan's frozen water could melt, briefly creating a warm, hospitable soup of organic compounds. When life is extinguished on Earth, Saturn's moon may be the home world of the Solar System's next life forms.

One chemical path leading to organic compounds on Titan appears in this simplified sequence. It begins 800 miles up as high-energy electrons (white) from Saturn's magnetosphere bombard a nitrogen molecule (tan).

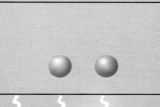

Energy absorbed by the nitrogen molecule breaks its molecular bond, freeing nitrogen atoms to combine with other atoms and molecules in Titan's atmosphere.

Meanwhile, solar ultraviolet radiation and cosmic rays (white) assault methane molecules, each made of one carbon atom (dark brown) and four hydrogen atoms (maroon).

Each methane molecule splits into two free hydrogen atoms and a separate fragment known as a methylene radical—made of a carbon atom and two hydrogen atoms (abbreviated CH_2).

Methylene radicals may combine to form a molecule of acetylene (C_2H_2, two carbon and two hydrogen atoms) or react with a free nitrogen atom to form hydrogen cyanide (HCN). Here, solar radiation strikes the hydrogen cyanide molecule.

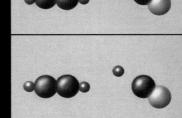

The hydrogen cyanide molecule splits into a free hydrogen atom and cyanogen (CN).

Acetylene joins with the cyanogen molecular fragment, resulting in a molecule of cyanoacetylene (HC_3N).

Cyanoacetylene and other complex organic molecules react with each other and with radicals, building long molecular chains called polymers. Here a polyacetylene chain ($C_{10}H_2$) begins its gradual descent to Titan's surface.

primitive Earth, to be sculpted by the magnetic field into a huge torus surrounding Saturn and its moons. Within that enormous structure, a smaller doughnut of energetic ions stripped from the icy surfaces of the smaller, inner moons—Rhea, Dione, Tethys, Enceladus, and Mimas—constitutes another, invisible ring around the Saturnian complex.

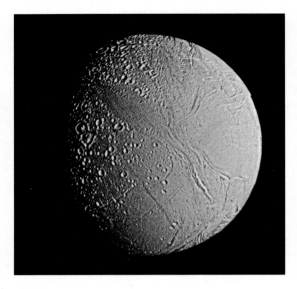

MOONS INTO RINGS

Until *Voyager 1*'s close-up, Titan held the title of largest satellite in the Solar System. Its 3,174-mile diameter turned out to be about 60 miles less than the diameter of Jupiter's Ganymede. But if the Voyager mission dropped Titan to second in size, it elevated it to first in scientific interest. As planetary astronomers had hoped, the probe's instruments sketched a world that might be a kind of planetary petri dish for the eventual emergence of rudimentary life forms *(pages 93-95)*.

All of Saturn's moons represent an entirely new class of planetary objects, vastly different from the Galilean satellites of Jupiter. The Jovian moons seem to mimic the distribution of material in the Solar System, with dense Io closest to the planet, and Callisto, a lightweight ball of rock and ice, farthest away. But Saturn's seventeen known moons are a chaotic jumble arranged in no particular sequence. Many of the smaller moons seem to be shards and fragments, remnants of collisions between larger bodies; such strikes may have supplied the first material for the rings. Some of the impact craters on the larger moons suggest that during the early history of the Solar System, Saturn's original complement of satellites may have been battered into the smaller stuff of rings by a rain of comets and asteroids. A so-called Deathstar crater on tiny Mimas strongly evokes that violent epoch. Measuring a third of the moon's 240-mile diameter, the crater marks an impact that must have come very close to cracking the little satellite apart.

In this scenario of a shattered early generation of Saturnian moons, gravitational attraction caused the resulting debris to reassemble into the present set of moons, with ice and rock mixed almost randomly. Gene Shoemaker, a planetary geologist with the U.S. Geological Survey in Flagstaff, Arizona, and a chief proponent of this somewhat controversial theory, believes it explains the differences between the Jovian and Saturnian systems. "Others see it all as unchanging," said Shoemaker. "My view of the solar system is dramatically different—that it's built of flying debris."

If this is true, the ringworld's geologic history must have been eventful indeed to produce the present collection of satellites.

Saturn's moon Enceladus, coated with unusually pure water ice, is the Solar System's most reflective body. According to one theory, the tidal stress resulting from the pull of Saturn's gravity heats the moon's interior and causes water geysers to erupt. The liquid then falls back to the surface, uncontaminated by rock and dust, to form a frozen, mirrorlike surface.

A crater sixty-two miles wide dominates the face of Mimas, one of Saturn's eight major satellites. The collision that produced the enormous pit—a third the width of the entire moon—was nearly fatal, but lesser craters, each the remnant of a separate impact, suggest that Mimas has survived eons of meteoric bombardment.

Clearly visible in this computer-enhanced image is the extended atmosphere of Titan, Saturn's largest moon. Sunlight scattered by suspended particles produces a bright orange arc more than a hundred miles above the Titanian surface; higher still, upper atmosphere haze and dust glow blue.

Saturn's major moons are mainly mid-size—their diameters are reckoned in hundreds, not thousands, of miles—and smaller than Jupiter's Galilean satellites; however, they are large enough to have undergone their own internal geologic convulsions.

Tethys, innermost of these, is about 650 miles in diameter and appears to be made mostly of ice. The moon boasts not only an impact crater, Odysseus, almost as wide as its radius, but also a trench known as Ithaca Chasma, 60 miles wide and 2 miles deep, extending at least three-quarters of the way around Tethys. Tiny Enceladus *(top),* just 300 miles in diameter, is covered with an icy surface that seems to have been melted and re-formed on a continuing basis, possibly by heat generated internally by the same kind of tidal forces that heat up the inside of Jupiter's fierce little satellite Io. Iapetus, farther from the planet, presents the faces of a cosmic yin and yang, one side a bright, icy hemisphere, the other nearly black. Scientists have yet to determine whether the geologic pigment of the dark side was produced inside the moon or introduced by an external source. Some believe the smudged hemisphere was caused by a dusting of material from Phoebe, another moon whose retrograde orbit suggests it was a visiting asteroid before being trapped by Saturn's gravity.

Many of these moons are coorbital, sharing an orbit in the same way that race horses share a track. The moons Janus and Epimetheus approach and recede from one another along the same path but never collide. Dione and Tethys each share a course with two small companions. Closer to the planet, the moons seem almost to shade into Saturn's luminous rings, the feature that, in the end, most compels the attention of science.

INFLUENCES SUBTLE AND PROFOUND

The very existence of the rings complicates every other line of study. For example, the shadow of the rings spreads over portions of the planet's upper atmosphere for years at a time, subtly shaping the atmospheric heat flow by intercepting the Sun's warmth. Similarly, interactions between the rings and the magnetosphere may influence the structure of each in ways not yet understood; the rings absorb subatomic particles so well that spacecraft readings taken under the rings suggest a condition rarely seen in the Solar System: an environment swept almost completely free of radiation.

The rings themselves occur in unbelievable profusion. Before the space age, three rings were known; today the number exceeds 100,000. At this level of detail, scientists have been able to confirm the existence of vast quantities of material in the ring plane, almost all of it millimeter-to-centimeter-size particles of water ice, with a few boulder-size objects thrown in. The rings maintain a sharpness of definition that still puzzles scientists, who now see an elaborate set of repeating encounters between moons and ring particles as a controlling factor in sustaining these sharp edges. A still-unexplained sorting mechanism also seems to be at work, segregating particles according to size into different portions of the rings.

The most persistent questions about Saturn's rings concern their origin and permanence. How did they get there, and how long have they existed? Scientists generally believe that the material in the main rings consists of the pulverized remains of satellites overtaken by what is called the Roche limit—the nearest orbit a satellite can have without being torn apart by the planet's gravitational forces. There is less agreement on the age of the rings. Before Voyager, scientists believed that the ring-building process occurred once in a planet's lifetime, very early in the history of the Solar System, and that the rings then remained essentially unchanged for billions of years. While many researchers still consider Saturn's rings to be as enduring as the planet itself, others have developed post-Voyager theories that envision dynamic, ephemeral rings. These come and go on a scale of tens of millions, not billions, of years as fragmented particles, eroded by repeated collisions, become electrically charged. Then, dragged into the planet's powerful magnetic field, they are swept downward into Saturn's atmosphere as ionized rain.

Whether they are permanent or not, the rings of Saturn seem to mimic processes that occur at much larger scales of size and motion, and in places light-years beyond the far planets. For example, the gravitational harmonics in which moons and rings interact at Saturn may involve spiral density waves—waves of compression and rarefaction that spread across the circular rings, concentrating and diluting the material there *(pages 102-103)*. Similar waves appear to be prime shapers of the vastly larger arms of spiral galaxies, in which the stars correspond to Saturn's ring particles. The echo of planetary processes in the greater dynamics of galaxies seems to say: This is a universe of ringworlds.

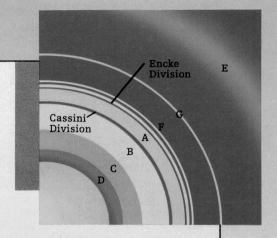

Encke Division

Cassini Division

E

G

F

A

B

C

D

A BEWITCHING RINGWORLD

Saturn's rings, a set of ribbons fashioned from un-countable icy fragments, rank among the chief wonders of the Solar System. The gleaming halo is enormous, extending 250,000 miles from an inner edge just above the planet's atmosphere to an outer rim almost too wispy to discern. It is also astonishingly thin, less than a hundred feet on average (although the diffuse rim spreads to a thickness of thousands of miles). The halo's features are the result of the gravitational interplay of Saturn, its moons, and the ring particles themselves. Most prominent are seven major bands *(above)*, lettered by astronomers in order of their discovery. Nearest to the planet is the dim, 4,000-mile-wide D ring. Beyond it lie three bright rings that form a nearly continuous sheet 40,000 miles in width: the dusky C ring, followed by the spoked B ring, and outside the dark Cassini Division, the A ring. Farther out, the spindly F ring is flanked by two small moons. Then comes the faint G ring and finally the E ring, 180,000 miles wide and thought to have formed only within the last several hundred thousand years.

In the past decade, space probes have revealed a wealth of structural variety: circular rings, spiral rings, clumpy rings, tangled rings. Drawing on data from the probes and theories they have inspired, the following pages explore the sixth planet's surrounding marvels.

A Stately Dance of Particles

Not even the intrepid *Voyager*s passed close enough to Saturn's rings to photograph an individual ring particle. But their findings, including radar studies and measurements of the rings' reflected light, confirmed scientists' long-standing conclusion that the bands consist entirely of fragments of material, probably resembling loosely packed snowballs of water ice. A slight dirty tinge implies small amounts of darker rocky material, and in some regions a reddish tint may signal traces of iron oxide—ordinary rust. Many of the ice chunks are the sizes of specks and pebbles, but some specimens may be as big as a house.

The fragments in all of the rings combined add up to no more than the mass of a single moon 200 miles across. The agent responsible for the scattering of this material is Saturn itself. Inside a boundary known as the Roche limit, powerful tidal forces resulting from the planet's gravity prevent particles from coalescing into large bodies, and they tear apart any entering moons more than 60 miles in diameter. (The size restriction varies with the body's density; a very low density moon could break up if its diameter exceeded even thirty-five feet.) Located roughly 45,500 miles from the top of Saturn's atmosphere, the Roche limit encompasses the inner rings—A, B, C, and D. Outside the Roche limit, rings F, G, and E are also entirely particulate, but they share their less stressful orbit space with the moons Mimas and Enceladus and several smaller satellites.

Because each ring fragment orbits Saturn independently, at a speed related to its distance from the planet, the rings revolve fluidly rather than as rigid hoops. Orbit times are short: For example, ice particles in the close-in C ring orbit in six to eight hours, while those of the A ring, 20,000 miles farther out, take from twelve to fourteen hours to complete a circuit. The particles frequently collide—jostling, nudging, rebounding, merging, and sometimes splintering in the course of their endless rounds.

Rings within Rings

In Saturn's A, B, and C rings—three neighbor bands less than 50,000 miles from the planet's surface—fragments are arranged into tens of thousands of ringlets and gaplets that give the effect of grooves in a phonograph record *(left)*. The structure is thought to result from two types of waves *(below)*, both products of the gravitational influence of Saturn's moons. The more powerful type is known as density waves, which are coiled patterns of compression and rarefaction that begin in zones where the orbit times of individual ring fragments are simple fractions of the orbital period of a nearby moon. In an effect known as resonance, the relationship between the two sets of orbits amplifies the moon's normally weak gravitational pull by providing repeated encounters with the same particles. The outcome is a dense pileup of fragments; the weak gravitational bonds between fragments enable the density concentration to pass outward like a ripple on a pond.

Mimas, the nearest large moon to Saturn, also stimulates the second type of wave. So-called bending waves corrugate the ring sheet into peaks and valleys thousands of feet steep, creating apparent ringlets and gaps in addition to those actually formed by the density waves. Bending waves arise because Mimas travels in an orbit that is slightly askew, taking the moon alternately above and below the plane of the rings and tugging particles up or down.

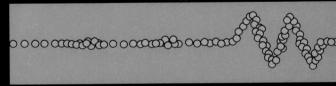

Shown in cross-section above are two types of gravitational waves that create grooves in the rings of the planet; both of these types are induced by Saturn's numerous moons. Density waves *(above, left)* travel horizontally, alternately gathering and dispersing particles into dense rings and sparse gaps. Bending waves, in contrast *(right)*, are like ocean rollers, perturbing the ring sheet into crests and troughs without affecting the density of the particles.

ELUSIVE SPOKES

ven the most complex patterns in Saturn's rings are
ssentially circular—with one exception. The maver-
ck is an ever-changing set of broad streaks, up to
,000 miles long, that extend like spokes on a wagon
wheel across the dense B ring *(right)*. The spokes
merge with stunning speed: A feature 4,000 miles
ong can form in just five minutes. Then, after rotating
or a few hours, they disperse. Given the dynamics of
he ring fragments—with inner fragments orbiting
aster than those farther out—this disintegration is
nderstandable. But astronomers are still unsure of
recisely how the spokes appear in the first place or
ow they persist for even a short time, despite the
ifference in rotation.

As each Voyager probe approached, the B ring's
narks appeared darker than their surroundings, yet in
unward photographs taken as the spacecraft reced-
d, the spokes seemed brighter—a light-scattering
ariation typical of fine dust. This led scientists to one
heory for the origin of the spokes. Like all tiny objects,
ust specks are particularly vulnerable to electrostat-
c effects, similar to the hair-raising static electricity
ncountered on Earth. Scientists speculate that the
nicroscopic particles must somehow acquire an elec-
rical charge, then lift from the ring sheet to interact
with Saturn's powerful magnetic field and create
isible markings.

The details of the process remain uncertain. Early
heories attributed the charge to sunlight or the ion-
zing effects of high-speed particles, but these sug-
estions have not held up well. A more recent proposal
that meteors spawn the spoke features when they
ollide with ring fragments. According to this theory,
uch encounters produce a plasma cloud of charged
ust and gas; as the charged matter orbits with the
ings, it intersects magnetic field lines passing per-
endicularly through the ring plane, which violently
edirect it over the rings. The rapid passage of the
lasma cloud stirs up ring dust, which forms a visible
poke along the plasma's trajectory. Over time, the
ust composing the spoke dissipates through orbital
otion and settles back into the B ring.

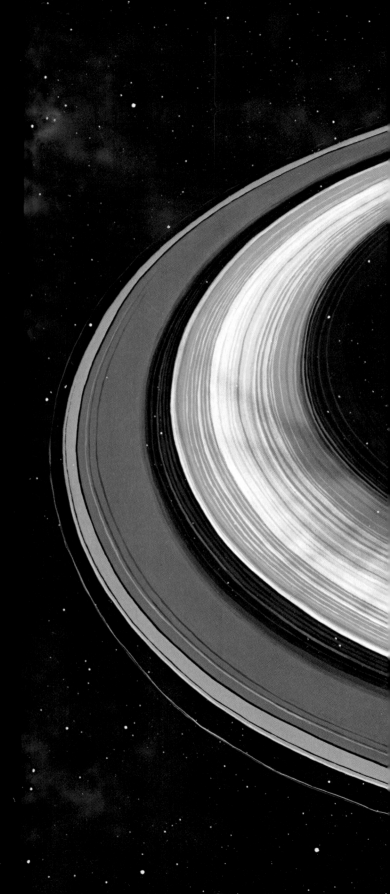

Shepherding Moons

Some 2,700 miles beyond the border of the A ring, seen as a series of white and brown lines in the near-edge-on view at right, lies one of Saturn's most intriguing features—the F ring, a slim band that seems to be made up of two or three interwoven strands. Each a few miles wide, the clumpy strands appear at times to overlap or cross; they are flanked by swaths of evenly distributed ice fragments, surrounded in turn by zones of microscopic dust. Dust included, the entire F ring system spans 300 miles, making it the narrowest of Saturn's assorted bands.

Two small moons, called shepherds, confine the ring particles to this straitened path: Pandora *(right, fore-ground)* on the ring's outer edge and Prometheus *(center)* on the inside. Though only fifty to eighty miles across, the moons together exert a gravitational scissor-lock that prevents the ring from spreading.

As illustrated below, the moons have slightly different orbital speeds, a key factor in the shepherding process. Prometheus orbits Saturn in well under fifteen hours, constantly overtaking the ring fragments. Pandora, for its part, circles the planet in just over fifteen hours, lagging behind the fragments. Every twenty-three days, the shepherds align, a position they are approaching at right.

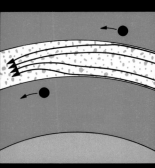

The F ring's narrow circlet of ice fragments is tended by a pair of shepherd moons, the inner one outpacing the fragments, the outer one moving more slowly. When fragments drift outward, the outer moon tugs at them from behind, reducing their speed so that they fall inward again. Conversely, the inner moon passes inward-straying fragments and pulls them forward, speeding them up and returning them to the F ring's thin stream.

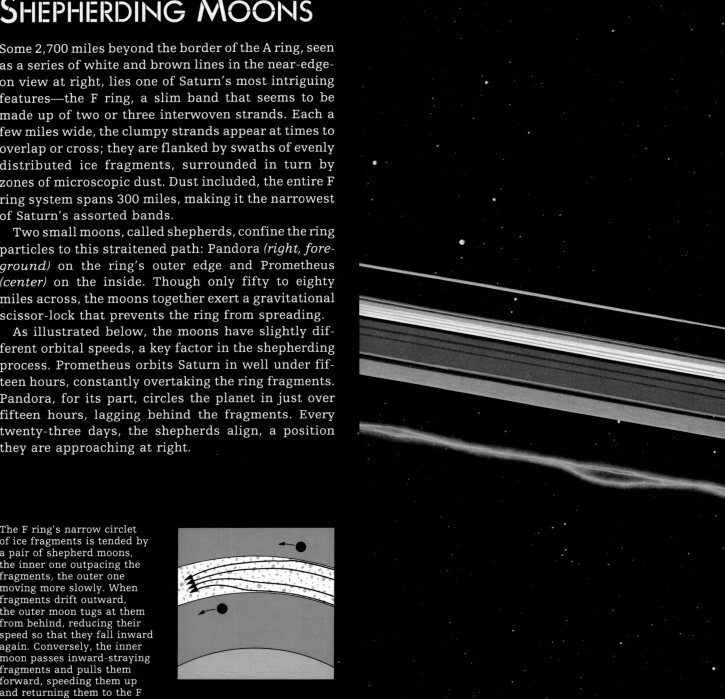

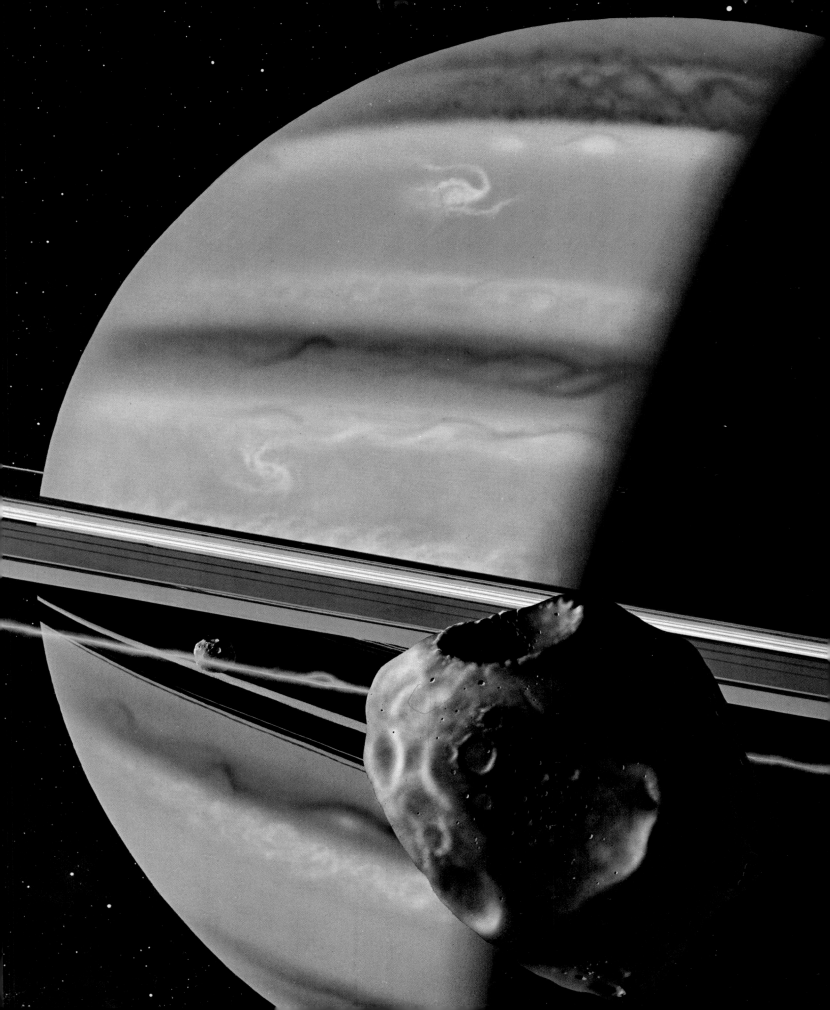

4/Outposts

Hanging like a blue-white firefly in this *Voyager 2* image, Uranus is attended by three of its five major moons, Ariel *(right)*, Miranda *(left)*, and Umbriel *(lower left)*.

fter its encounter with Saturn in late August 1981, *Voyager 2* began what would be four and a half years of coasting through the outback of the Solar System in pursuit of its next target: Uranus. Saturn shrank to a bright point of light in the probe's wake, and the solar wind, which had blown strong at the fifth and sixth planets, dwindled to an electrified zephyr.

By now the spacecraft was beginning to feel its age—and the age of the technology that created it. Initially given only a 60 to 70 percent chance of lasting five years, *Voyager 2* steadfastly went beyond all expectations. The gritty little ship, already four years old at Saturn, would be eight at Uranus and, if the mission continued as planned, twelve when it reconnoitered Neptune. Come that encounter, children in first grade when the probe lifted off would be entering college.

The craft's vital signs had weakened. Its voice was barely audible at Earth now, and the plutonium oxide fuel supply had dwindled. The 470 watts available at launch had dropped to 400; a full encounter mission, as thorough as those conducted at Jupiter and Saturn, was now more than *Voyager 2* could deliver. The low light levels this far from the Sun required long exposures that the probe's motion would smear into illegibility. And the camera platform that had malfunctioned in Saturn's shadow remained frail.

Voyager 2's larger problem was that, like *Pioneer 11* earlier, it was taking on jobs for which it had not been designed and which had not even been added until the spacecraft, having swept past Jupiter, was en route to Saturn. But the scientific motivation behind the new assignments was compelling. The planetary alignment that had prompted NASA to explore the far planets would not repeat itself for nearly two centuries, and no follow-on missions beyond Saturn were even on the drawing boards. *Voyager 2* offered the only close look humanity would have of this distant realm for many years to come.

Before Voyager, what little had been deciphered of these planetary outposts spoke of violent events in the dim past of the Solar System, when whole worlds, it seemed, had been bludgeoned and reversed in their tracks or hurled into crazily eccentric orbits. Telescopes showed Uranus as a small, featureless, blue-green disk whose spectrum and moon orbits indicated that the planet had somehow been tipped on its side. Neptune, visually a near twin of

Uranus, offered still less to terrestrial observers beyond the hint of primeval disturbances preserved in the troubled orbits of its two known satellites. As for unattainable Pluto, it was so distant that it could not even be resolved into a planetary disk. Beyond there, all was mystery.

In Pasadena, listening to their ailing, aging robot at the end of a radio link more than a billion miles long, Voyager's masters took what amounted to a daring medical decision. If *Voyager 2* was to be of any use at all in the outer reaches of the Solar System, the team would have to operate.

SURGERY AT A DISTANCE

The first order of business for the engineers at the Jet Propulsion Laboratory was to find out whether the platform bearing the cameras would work at Uranus. Because the platform had begun to stick near the end of the Saturn encounter, the engineers were wary of tampering with it unnecessarily. Instead, they built eighty-six full-scale replicas of the misbehaving assembly, a jammed gear train in a motorlike device that translated an electrical signal into mechanical motion. These laboratory specimens were subjected to a battery of rigorous tests that included putting an identical copy of the errant gearbox through a simulated mission. One of the test gears stuck after 348 operations of the scan platform. *Voyager 2*'s had stuck at 352. The laboratory post mortem showed that the gears were inadequately lubricated. Used too often and too fast, they tended to grind together and eventually to seize. Further experiments revealed that by alternately operating the gears and giving them a rest to cool them down, the platform could be freed.

Then the engineers rehearsed their newly developed stratagems on a spacecraft in flight, using *Voyager 1* as a stand-in for its ailing twin. When they were satisfied that their commands would not aggravate the problem, they ordered *Voyager 2* to begin gently "sluing" its scan platform by moving it back and forth horizontally. In February 1983, the platform was fully functional once more, but under a new rule: Run a little slower.

The next step was to determine whether *Voyager 2* could be made to compensate for the expected low light levels. Sunlight is four times weaker at Uranus than at Saturn and almost two and a half times weaker again at Neptune. The problem would be to keep the cameras steady during the necessary long exposures, so that the ship's motion would not destroy a generation's only close-up images of these planetary systems. The solution was what JPL's Howard Marderness called an "antismear campaign" against three sources of image-wrecking motion. Marderness's strategy carried risks of its own. JPL engineers would have to go into the in-flight computer programs managing the spacecraft, where a wrong move could send *Voyager 2* crazily out of control.

In the virtually friction-free environment of space, every action, however small, produces a reaction. Aboard *Voyager 2,* this meant that a slight nodding motion was imparted to the spacecraft every time its recorder switched on or off. Though minuscule, the movement was enough to blur close-up images.

This difficulty was addressed by developing a new program for the thrusters that controlled the craft's orientation. The thrusters would be instructed to fire short bursts whenever the recorder started or stopped, thus counteracting the unwanted motion.

Trouble also came from *Voyager 2*'s persistent swinging back and forth in relation to its exact course heading. Like the helmsman of a ship, the space probe's computers applied corrections by ordering the thrusters to fire short bursts that brought it back in line with its heading and kept its radio antenna from swinging too far out of alignment with Earth. Though only ten milliseconds in duration, each firing imparted a faint wobble to the craft. JPL engineers, worried that even ten milliseconds was too long to ensure a steady camera platform at Uranus and Neptune, opted for five-millisecond bursts on the approach to Uranus and four-millisecond bursts at Neptune.

Shorter firings also meant more firings, which raised another specter: Increased use might wear out the thrusters before their time. Turning to the manufacturers of the devices, JPL arranged a lengthy series of tests between October 1984 and April 1985. Engineers fired the test thrusters 25,000 times for different durations—far greater use than the mechanisms would ever get on *Voyager 2*—and the devices held up fine. Having established the durability of the thrusters, the engineers once again used *Voyager 1* as a flying laboratory, this time to develop the techniques they would need to reprogram the computers on its counterpart.

A third sort of motion difficulty was presented by *Voyager 2*'s cruising speed. The probe would go by Uranus at about 40,000 miles per hour, a velocity that would ruin high-resolution shots taken at closest approach, 50,000 miles above the planet's cloud tops. Here, the answer was to make the entire spacecraft turn to pan with the cameras, compensating for the blurring effects of flyby speed the way one would stop a fast-moving subject by tracking it with a camera. Once this panning tactic had been programmed into *Voyager 2*, the craft would be ready for the Uranus shoot.

With *Voyager 2* only months from Uranus, the engineers took the fateful leap and cracked open the ship's flight-control programs. "Testing computers on board a spacecraft is like open-heart surgery," commented flight engineering manager William McLaughlin. "If anything goes wrong, you haven't got a second chance." In this case, the operation went smoothly and was completed by the time *Voyager 2* began its fall toward Uranus.

Throughout their long-range endeavor, mission managers had been faced with another, more basic problem. No matter how clear and sharp the Uranian images turned out to be, they would be worthless if the Voyager probe could not send them back to Earth. Enfeebled by the 1.9 billion miles they would have to travel from Uranus, the spacecraft's radio signals would arrive at tracking stations on Earth with a strength of just one ten-million-billionth of a watt—so faint that if the energy were collected for more than three billion years, it would barely keep a refrigerator light aglow for one second. The faintness of *Voyager 2*'s signal would make its rapid chatter in the radio X-band difficult to distinguish from the snaps and crackles of the ever-present background noise in the transmission. To be intelligible, the probe would have to speak in a kind of digital drawl. Here, an early precaution paid off.

Before launch, mission engineers had added a speed-boosting backup feature to the normally poky S-band radio unit to allow the housekeeping channel to carry images and other data if the faster X-band radios failed. Now, with *Voyager 2* heading toward Uranus and Neptune, JPL controllers remembered the dormant system and began gingerly to wake it up. By switching the backup feature to the X-band radios, mission controllers could use it to speed up the transmission of slowly enunciated messages. Mission engineers also rewrote the probe's imaging software, reducing by almost two-thirds the number of bits needed to transmit a given picture. By January 18, 1986, *Voyager 2* was sending back early views of Uranus from about six million miles away, flying a near-perfect trajectory. After an eight-year journey of three billion miles, the probe was within 100 miles of its designated path, the navigational equivalent of sinking a thousand-mile putt.

Then, just as it seemed the encounter would be trouble-free, *Voyager 2* developed some alarming new symptoms. On the monitors at JPL, images of Uranus began suddenly to streak with blank lines. With less than a week to go until the time of nearest approach, engineers began a frantic search for the source of this inexplicable plague of holes in the spaceship's messages.

"I was firmly convinced it was somewhere in the ground system," McLaughlin recalled, referring to the Deep Space Network used to communicate with the Voyager craft. In the course of nearly four years, the global array of huge antennas had undergone a $100-million upgrade of its electronics and tracking equipment to help it deal with ever-fainter signals. Despite the improvements, the network had not quite regained the full confidence of the mission staff. But a nail-biting weekend search of the array uncovered no obvious faults.

Suspicion next fell on the spacecraft itself—in particular, on what one mission scientist had called its occasionally "psychotic" computers. As a form of medical checkup, JPL commanded *Voyager 2* to transmit to Earth the entire contents of the science computer's memory—all the software used to process information from the probe's sensors. After poring through it digit by digit, the engineers were at last able to pinpoint the error: In the torrent of zeros and ones constituting the electronic expertise of Voyager's computers, a single bit had flipped from zero to one. A program fix to sidestep the failed bit was quickly devised and beamed up. Three days before its closest encounter with Uranus, *Voyager 2*'s pictures were once again flawless.

Thanks to its astrophysicians in Pasadena, the ship that entered the outskirts of the Uranian system was a distinct improvement on the one that had left Earth nearly a decade earlier. And now it had a brief appointment with a world humanity had barely seen.

Voyager 2's encounters with Jupiter and Saturn had been a leisurely stroll, with the probe steering for a day and a half among moons and rings. Its visit to Uranus entailed a different dimension of speed. To the advancing spacecraft, the rings of Uranus and the orbits of its moons were like the concentric circles of a target: The planet itself was at the bull's-eye, and *Voyager 2* was a bullet whizzing toward it at eleven miles a second. The ship would have just six hours to make all its high-resolution observations of the entire system. Moreover, with radio transmissions taking two and three-quarter hours to reach Earth, there would be no time to ask help of Pasadena. *Voyager 2* would have to rely solely on the instructions it had already been given. The probe was on its own.

BLUE IMAGININGS

At Jupiter and Saturn, *Voyager 2* had glimpsed many-hued bands, spots, and swirling plumes, even from a distance. But the aquamarine face of Uranus, as wide as four Earths, grew steadily before the ship's camera eyes without revealing the slightest atmospheric detail. Andrew Ingersoll, the planetary weatherman for the two previous encounters, later recalled that, for a time, the imaging team was called the "imagining team." As one disappointed JPL scientist grumbled, Uranus was little more than a "fuzzy blue tennis ball."

The cause of that blandness was a thin haze of methane ice crystals and various particulates that enveloped the planet and drew a veil over any turmoil in the atmosphere below. Researchers later theorized that the Uranian atmosphere, many thousands of miles thick, may be a chilled, super-dense brew of hydrogen, helium, and water surrounding a solid core not much bigger than the Earth. By absorbing only red light and reflecting all the rest, the methane in the atmosphere colored Uranus greenish blue.

Still, the veil was not totally impenetrable. Through the wizardry of image processing, the JPL team managed to tease out of the Voyager pictures faint details in the Uranian atmosphere even though those details were only about five percent brighter than their dull surroundings. For example, the south polar region, facing the Sun, appeared to be smudged with a reddish tinge, probably the result of simple organic chemicals born of the reaction between methane and sunlight—a process not unlike that which creates photochemical smog on Earth. Even more interesting, a series of concentric bands could be clearly seen on the computer-massaged images, and bright wisps of high-flying clouds were detected in the aquamarine atmosphere. The clouds were apparently whipped along by bands of powerful east and west winds with speeds up to 350 miles an hour. Because Uranus crabs around its orbit on its side, with its poles aimed at the Sun for decades at a time, scientists expected to find correspondingly bizarre wind patterns on the planet, governed by the peculiar way in which solar radiation percolates through the Uranian atmosphere. Instead, they found bands of prevailing winds resembling those seen

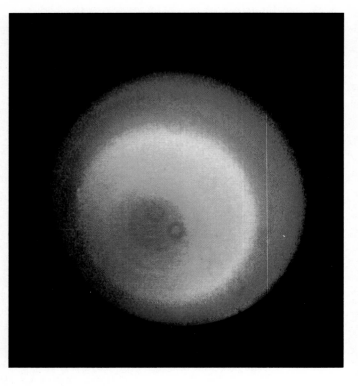

Subtle atmospheric differences not seen in ordinary photographs of Uranus emerge in a composite computerized image of the planet in both visible and ultraviolet light. At center, a brownish haze covers the sunward pole; scientists theorize it is caused by acetylene, which forms as direct sunlight strikes methane in the atmosphere. Around the pole, a series of bands imply cloud and wind patterns similar to those seen on Jupiter; their concentric arrangement suggests that the winds result from Uranus's rotation.

in the atmospheres of other cloud-covered planets and a surprising uniformity in temperatures from pole to pole. Clearly, the planet's spin, not solar radiation, was the dominant force shaping the large-scale circulation of Uranus's atmosphere—and, presumably, the atmosphere of any planet. Still two weeks and 14 million miles from Uranus, *Voyager 2* had taken a sizable swipe at meteorology's conventional wisdom.

Having wind bands and a few Uranian clouds in hand eased the gloom among the "imagining" team, but there was little to cheer the particles and fields group at Pasadena. Prepared by their Saturn experience to find a relatively compact magnetosphere at Uranus, the JPL team began to suspect Uranus had no magnetic field at all. As late as mid-January 1986, with *Voyager 2* only a few days from Uranus, instruments had failed to report evidence of magnetism. If there was no field at Uranus, a vital window would be closed to science. The JPL scientists had counted on a magnetosphere— which would spin with the planet—to indicate the internal rotation speed of Uranus. Without one, they would have no frame of reference for measuring atmospheric motions across the world's almost featureless face.

Then, on January 19, just five days and 4.5 million miles from Uranus, the spacecraft detected radio emissions, clear signs of synchrotron radiation generated by the movement of high-speed electrons in a magnetic field—but the Uranian magnetosphere itself was still far away. *Voyager 2* did not cross the telltale bow-shock region between the magnetosphere and the solar wind until about ten hours before its closest approach on January 24.

The magnetosphere penetrated by the craft rises only 370,000 miles or so above Uranus on the sunward side, tailing out behind some 3.7 million miles on the night side. Flapping gently in the solar wind, the magnetic field twists into corkscrew patterns, like the submerged wake of a ship's propellers. But it carries a powerful sting in the form of trapped radiation. This may have left its mark on the planet's moons and slender rings. Some scientists believe the distinctive darkness of the rings and small satellites may be a kind of scorching, caused as methane on their surfaces was bombarded over the millennia by trapped high-energy particles.

Like almost everything about the Uranian world, the magnetic field was not what previous encounters had taught the scientists to expect. In contrast to other planets with magnetic fields, where magnetic and rotational axes match to within a few degrees (Saturn's are almost perfectly aligned), the Uranian field turned out to be about sixty degrees out of alignment with the axis of rotation. This odd arrangement led some scientists to speculate that the distant world was undergoing a gradual reversal of its magnetic poles.

Lopsided as it was, the field made a useful clock. The planetary radio astronomy team, headed by James Warwick of Colorado, measured radio emissions generated by the magnetic field as its off-center axis spun around the planet's axis of rotation. By observing periodic fluctuations in the radio signal, the group could recalculate the length of Uranus's day: 17.24 hours, up from the estimate of 16 hours derived from ground-based observations.

BRAVE NEW WORLDS

As was the case in *Voyager 2*'s other encounters, much of the mystery of Uranus resides in its entourage of satellites. Five moons were known as points of light before the Voyager visit. During the probe's plunge toward Uranus, however, nine tiny undiscovered moons popped out of the darkness between the planet and the orbit of Miranda, the innermost of the known satellites. A tenth new moon was found in this lunar herd later in the encounter. But it was the old moons that told the most enthralling tale.

Before the rendezvous, geologist Laurence Soderblom of the Voyager imaging team had predicted the Uranian satellites would show "impact craters and little else." Oberon, first of the so-called Big Five moons the probe would scrutinize and the most distant from the planet, was pocked indeed. But the floors of some of its craters turned out to be flooded with dark material. In the remote past, Oberon may have been the scene of localized volcanic eruptions, spewing out a warm ice-rock slush from its interior.

Signs of an even more action-packed youth were evident on Titania, next moon in toward the planet. At just under a thousand miles in diameter, Titania is a twin to Oberon in size if not in appearance. Here *Voyager 2*'s cameras saw more than craters. Great valleys hundreds of miles long wrinkled the moon's surface. These huge, winding trenches may have formed when water inside Titania froze and expanded, cracking the moon's thin, stiff crust.

Umbriel, middle member of the five, seemed to hold no surprises. Dark and drab, it showed little sign of present or past geologic activity beyond intensive cratering of its surface. But even this unprepossessing place had its puzzle. As *Voyager 2* drew to within a quarter of a million miles of the moon, it detected among the countless indentations a single white ring. Later dubbed the "fluorescent Cheerio," the ring lay inside a large crater and appeared to consist of ice kicked up from Umbriel's cold interior. The question for the scientists in Pasadena was less what the Cheerio was than why the moon had only one. This has remained Umbriel's secret.

Ariel, the next moon in, displayed a sinuous network of smooth-floored valleys reminiscent of rift valleys on Earth. These were almost certainly formed by a stretching and cracking of Ariel's crust billions of years ago, followed by the upwelling of warm material through splits in the valley floors. Clearly, the moon had endured a tumultuous geologic childhood.

Like all the satellites of Uranus, 300-mile-wide Miranda had promised to be no more than a cratered ball of ice. But the moon revealed itself to *Voyager 2* as one of the most geologically complex bodies in the Solar System, a montage of natural disasters. Parts of Miranda had been torn and wrenched into stupendous cliffs, canyons, and ridges. In one place, a precipice soared up for twelve miles—more than double the height of Mount Everest. Strangest of all were two huge, almost rectangular regions resembling monstrous racetracks, and another, chevron-shaped feature. The moonscape bounding them is scored with ridges, as if furrowed by a giant plow.

To explain the origin of such amazing terrain, geologists on Earth put forward various exotic theories, none of them conclusive. According to one, Miranda froze at the height of its early turmoil, eternally preserving features that had long since vanished on every other world. A more dramatic possibility is that the little moon was shattered, perhaps as many as five times, by collisions with asteroids. Miranda's fragments, having remained close together, then gradually reassembled as a jumble of ice and rocky chunks.

Miranda and its neighboring moons presented scientists with a geologic enigma. "The bodies themselves are so low mass," explained Bradford Smith, head of Voyager's imaging team, "that it would be hard to get any melting in the core. And yet there is very clear evidence for some kind of internal activity on all but Umbriel. Miranda is so bizarre we haven't come up with any good explanation. And why is Umbriel so dark? Every time we think we understand these planets, we discover when we get there that we don't."

The moons are only part of the larger Uranian mystery. Some scientists hypothesize that before the satellites existed in their present form, a cataclysm smashed Uranus onto its side. According to some calculations, an Earth-size object, traveling at 40,000 miles per hour—roughly *Voyager 2*'s flyby speed—may have been the culprit. Once bowled over by the intruder, Uranus may have cast off a huge, hot disk of gas that then cooled and coalesced into the family of satellites seen today.

By contrast, the Uranian rings appear to be more recent. Nine were known from Earth-based observations prior to the Voyager visit, and the probe discovered two more. But it was what the craft failed to find that was most intriguing. Contrary to expectations, the strands around Uranus proved to be relatively dust free, composed instead of objects almost exclusively bigger than a suitcase and blacker than coal. By some process still to be elucidated, the rings have been swept clear of all smaller particles.

Voyager 2 found only two tiny shepherd moons, fewer than had been anticipated. Without the gentle gravitational influence of such moons, which play a key role in

Ridges, craters, canyons, escarpments, and a gigantic chevron-shaped formation *(center)* bear witness to the cataclysmic past of Miranda, smallest of Uranus's major moons. Researchers theorize that the satellite shattered billions of years ago in a collision with planetary debris. Attracted by their mutual gravity, the fragments later regrouped but were too cold to fully meld, leaving Miranda with a tortured surface.

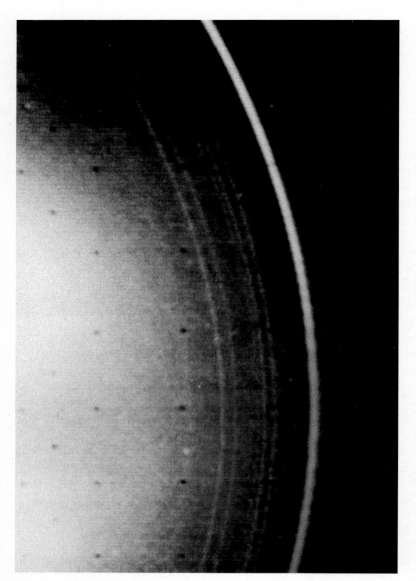

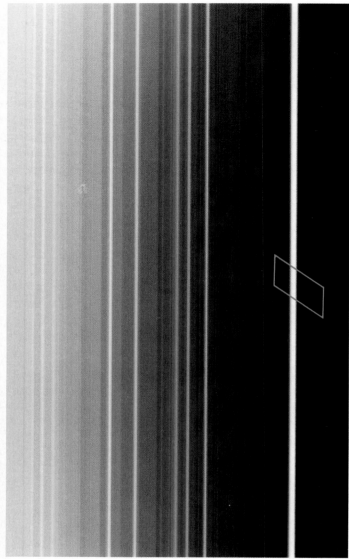

containing the rings of Saturn, scientists were hard pressed to explain how the Uranian rings could remain sharp edged and narrow—most are less than ten miles wide. One possibility: The rings were formed quite recently by the breakup of a small inner moon. If such is the case, they could be short-lived, the material in them doomed to tumble into the planet's bland, blue-green atmosphere over the next few million years.

Virtually everything about the Uranian system—from its inexplicably slender rings to its cockeyed magnetic field—goes against scientific experience. "To create a historical scenario for what Voyager saw at Uranus, we need more miracles than any thinking person will accept," Brad Smith said after the encounter. By then, *Voyager 2* had shifted course, heading for Neptune.

THE DEEP

In an orbit one and a half times farther than Uranus from the Sun, Neptune chugs sluggishly through the void. A full Neptunian year—165 Earth years—

In 1986 *Voyager 2* images like these gave astronomers a first clear look at Uranus's slim rings. At left, a picture taken from two and a half million miles away includes the planet and its ten strands. Above, variations in ring color show up in a closer, computer-enhanced image, suggesting chemical variations on a basic rock-ice makeup. At right, a twenty-mile-wide close-up of Uranus's brightest ring shows very narrow bands sighted as a star passing behind the ring dimmed and brightened. Yellow and white indicate dense regions that blocked starlight, and red represents relatively empty zones through which the star shone.

118

has not elapsed since the planet's discovery in 1846, and will not until 2011. Despite its remoteness, Neptune, by the time of the Voyager probes, had begun to reveal that it was not merely a slightly smaller, slightly more massive version of Uranus. This other blue-green subgiant showed promise of being a highly individual, various world in its own right.

Under ideal seeing conditions, ground-based telescopes tuned to the near infrared could detect detail in the eighth planet's distant atmosphere, where weather patterns appeared to change over hours or days and to alter seasonally. Neptune looked more mottled and banded than Uranus and seemed in some respects less finished, continuing to bring heat up from its interior.

Spectroscopic observations indicated that its atmosphere is about 97 percent hydrogen and helium, with about the same amount of methane as Uranus. Many scientists believe that Uranus and Neptune represent intermediate steps in evolution between the dense inner globes and the gas giants, Jupiter and Saturn. Both contain larger proportions of elements heavier than helium; calculations of Neptune's mass indicate that of the four gaseous worlds, its endowment of such elements is greatest.

Like Uranus, Neptune appears to be the victim of violence, at least to judge by the orbits of its two known satellites. Nereid, an estimated 500 miles wide and the smaller of the two, follows a wildly eccentric path, swinging to within 800,000 miles of its home world and then back out some six million miles into space. Triton, about fifteen times broader than Nereid, is a virtual planet, possessing a nitrogen-methane atmosphere and covered, some scientists surmise, with freezing seas of liquid nitrogen dotted with methane icebergs. But the moon is as ephemeral, in cosmic terms, as a moth circling a flame. Triton follows a track opposite its planet's direction of spin, the only retrograde orbit among the Solar System's major moons. Because of tidal forces arising from that reverse motion, Triton is being made to spiral slowly in toward Neptune. In about a hundred million years, it will be torn apart by its planet's gravity and sculpted into rings that may rival Saturn's.

One theory holds that Triton and Nereid are simply captured asteroids. But another, very different scenario involves both moons in a fantastic game of primordial interplanetary billiards. According to this view, some large object, perhaps as massive as five Earths, blundered through Neptune's brood of satellites millions of years ago. The gravitational effect of its passage was so traumatic that not only were the paths of Triton and Nereid disrupted, but a third moon of Neptune may have been thrown out of its old orbit altogether to become a ninth planet: Pluto.

Neptune's peculiar moons seem to be matched by an odd, incomplete set of narrow rings made up of segments of a circle and orbiting the planet at altitudes of some 10,000 to 27,000 miles. The presumed arcs were detected in occultation experiments—ultraprecise measurements of the interruption of light from a star as Neptune passed in front of it. On the basis of the way the starlight flickered during the observation, scientists concluded that there must be an intermittent ring around the planet, held in place by Neptune's gravity or perhaps by the confining action of another, still-undetected moon in a highly inclined orbit.

To avoid these rings and hypothetical moons, *Voyager 2* was sent a command in March 1987 to fire its thrusters for seventy minutes and thirty seconds, using precious fuel to gain an additional 20.5 miles per hour. This midcourse maneuver also altered the craft's trajectory to advance its arrival time at Neptune by about twelve hours, when Australian radio antennas would be better positioned to receive the probe's faint messages. The adjustment aimed *Voyager 2* at a point in space where, on August 24, 1989, at 9:00 p.m. Pasadena time, it should whip past Neptune just 3,100 miles over the planet's north pole, narrowly skirting the arc rings and radiation that scientists believe are trapped in the magnetic field. Five hours later, the spaceship is scheduled to pass within 25,000 miles of Triton and then begin its eternally long haul toward interstellar space.

Beyond Neptune, *Voyager 2* will ultimately find only the faint physical boundary where the attenuated solar wind gives way to winds from other stars. The discovery of that border will mark the end of the Voyager epoch of planetary exploration. Pluto, the last known world in the Solar System, must await another pathfinder.

THE ODD COUPLE

An average 3.66 billion miles from the Sun, the orbit of Pluto threads through a realm as forbidding as the mythic hell of its namesake god. Here, the Sun's average radiation is some fifteen hundred times less than it is at Earth, a kind of moonlight falling through a thin atmosphere upon frozen sheets of methane and water ice. Pluto's day is about an Earth week long, its year just under 248 Earth years. The ninth planet will not return to the orbital point where it was discovered until the year 2177.

Where the other planets' orbits lie no more than seven degrees out of the plane of the ecliptic—the approximate plane of the Solar System—Pluto's is tipped more than seventeen degrees. This high inclination is matched by a

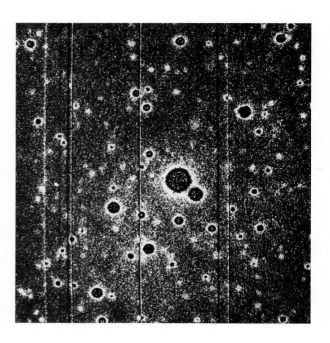

Dozens of Milky Way stars throng an electronically recorded image of Neptune *(center)* that was obtained in 1984 with the 100-inch telescope at Las Campanas Observatory in Chile. Astronomers were able to discern several high-altitude clouds on the planet's face, features that allowed them for the first time to measure the length of Neptune's day—a period seventeen hours and fifty minutes long.

A computerized portrait of Neptune's moon Triton depicts the body as NASA imaging specialists expected it to appear to the cameras aboard *Voyager 2.* Its slightly reddish cast comes from surface hydrocarbon sludge, while formations ranging from nitrogen-methane polar caps to dark, liquid nitrogen lakes might result from the moon's temperature of minus 350 degrees Fahrenheit.

120

wild orbital eccentricity that takes the tiny world nearly four and a half billion miles away from the Sun, then sweeps it back to within three billion. From 1979 to 1999, Pluto's orbit is closer to the Sun than Neptune's.

Because the ninth planet is an unresolved, starlike point of light to even the largest ground-based telescopes, astronomers do not know its precise size, and they have repeatedly revised their estimates. Once reckoned to have a diameter of about 3,600 miles, Pluto was shaved down to 1,370 miles by Infrared Astronomical Satellite (IRAS) observations analyzed in 1986. A year later astronomers at the University of Hawaii inflated it to a slightly more robust diameter of 1,504 miles—an upgrading that still left the little body with only a slim claim to planethood. Its status as a planet rests today almost entirely on the fact that it has an atmosphere and a moon. But what a moon.

Named for the grim boatman who ferries souls across the river Styx to the god Pluto's dismal empire, Charon is in some ways without peer in the Solar System. The satellite was discovered in 1978 by James Christy of the U.S. Naval Observatory, on photographs taken by the navy's sixty-one-inch reflector in Flagstaff, Arizona. (The instrument is located a mere four miles from the telescope Clyde Tombaugh used to find Pluto a half-century before.) Christy noted that some images of the planet were elongated by a kind of bump, suggesting the periodic appearance of a second object nearby.

This object turned out to have a diameter about half that of Pluto. Charon is, in fact, the smaller half of what amounts to a double planet. Evidently these

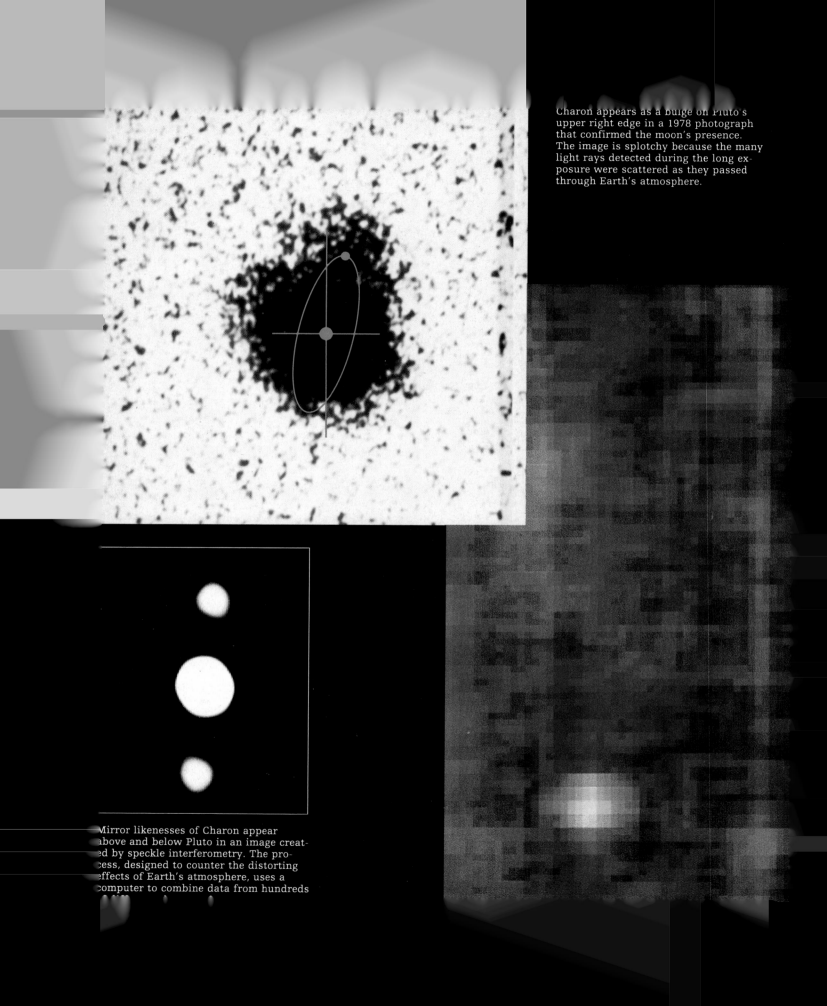

Charon appears as a bulge on Pluto's upper right edge in a 1978 photograph that confirmed the moon's presence. The image is splotchy because the many light rays detected during the long exposure were scattered as they passed through Earth's atmosphere.

Mirror likenesses of Charon appear above and below Pluto in an image created by speckle interferometry. The process, designed to counter the distorting effects of Earth's atmosphere, uses a computer to combine data from hundreds

PROBING PLUTO AND ITS MOON

Astronomers must reach deep into their bag of tricks to discover even the most general facts about the farthest-ranging planet of the Solar System. Smaller than Earth's moon and more than 10,000 times as distant, Pluto and its satellite Charon together appear as no more than a point of light when viewed by the most powerful earthbound telescopes. Charon, almost half the size of the planet, went unnoticed for nearly fifty years after Pluto's discovery.

Scientists have begun to penetrate the Plutonian shroud of mystery by using new methods to wring more data from the dim sunlight reflected by the two bodies. Photographic plates that had been exposed for a very long period provided the first evidence of Charon's existence. Multiple short-exposure images of the Pluto-Charon system were analyzed by computers in astronomers' efforts to determine the size and sepa-

ration of the two partners, and infrared data gathered by a satellite in the Earth's orbit supplied information about Pluto's composition.

The planet's origin has also become the subject of computer analysis. Pluto was once thought to be a former moon of Neptune, its path through space drastically altered by interaction with Neptune's largest moon or some other massive object. Either of these possibilities, however, would require an unlikely sequence of events; most scientists now think that Pluto was a planet from the start, perhaps originally orbiting in a near-circle like other members of the solar family. With the help of powerful computers, astronomers are trying to determine how such a path might have naturally evolved to the planet's present eccentric and inclined orbit over the 4.6-billion-year life of the Solar System.

Pluto and Charon appear within a yellow-green region in a false-color image from NASA's Infrared Astronomical Satellite *(left)*. From infrared data, scientists deduced that Pluto has methane ice caps, a warmer equatorial belt, and a rarefied methane atmosphere 870 times thinner than the Earth's.

THE PUZZLE OF CHARON

The most twinlike of any planet-moon pair in the Solar System, Pluto and Charon are enigmatic partners. Many scientists think that the two bodies were formed like other worlds and their satellites, slowly accreting from a swirling cloud of gas and dust and remaining gravitationally bound to each other in a double-planet system *(top right)*. Others suggest that a large object speeding through space collided with Pluto, knocking off a chunk that became its moon *(bottom right)*. Less likely is the theoretical possibility that Pluto's gravitational field pulled a passing planetesimal into orbit around the planet.

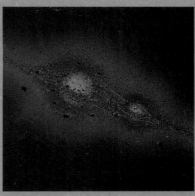

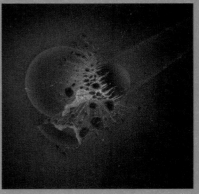

FOCUSING ON A
SPECK OF LIGHT

Scientists have developed remarkably detailed models of Pluto's surface and spin by studying the colors and brightness of the planet's light. During the early 1950s, astronomers discovered that the intensity of the light rose and fell in a recurring pattern. This observation was soon explained as a consequence of Pluto's rotation: If the planet's surface had some regions that were reflecting light more efficiently than were others, then Pluto would appear brighter as it turned those regions toward the Sun. When plotted on a graph like the one below, the cycle was found to re-

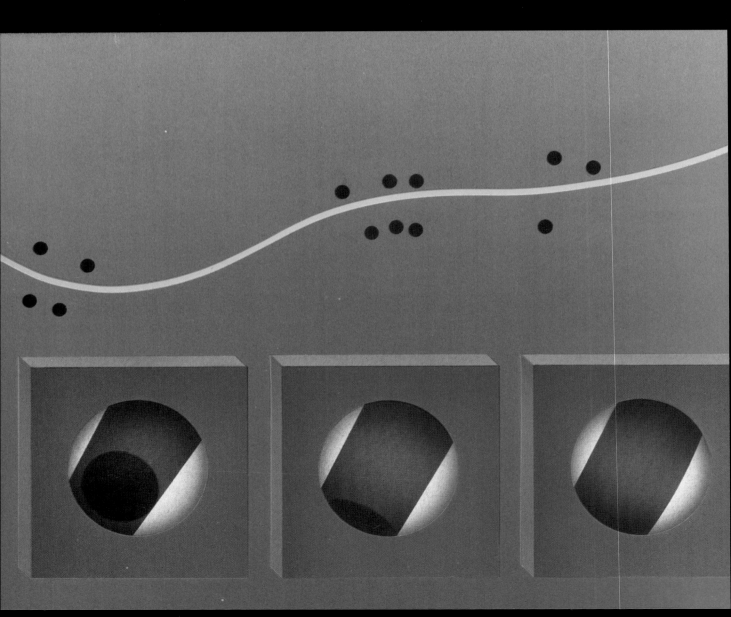

peat every 6.4 days—the period of a single revolution.

The brightness measurements provided a basis for rough black-and-white representations of Pluto's surface like the one at bottom, shown revolving in coordination with the graph. This model is augmented with colors based on spectroscopic studies, which suggest that methane ice covers much of Pluto's surface. A muted reddish tint indicates that rock may cover some of the rest of it.

New insights sprang from the observation that Pluto's light became progressively dimmer from the 1950s to the 1980s, even though the planet's eccentric orbit was bringing it closer to the Sun. Scientists proposed that the dimming was the result of substantial tilting of Pluto's axis of revolution. If that axis was not perpendicular to the planet's orbital plane, Pluto would present various regions to the Sun at different points on its 248-year orbit. This slowly changing view of the regions could cause the long-term changes in Pluto's brightness. Further studies showed that the tilt was sixty-two degrees from the perpendicular, enough to account for the observed fluctuations.

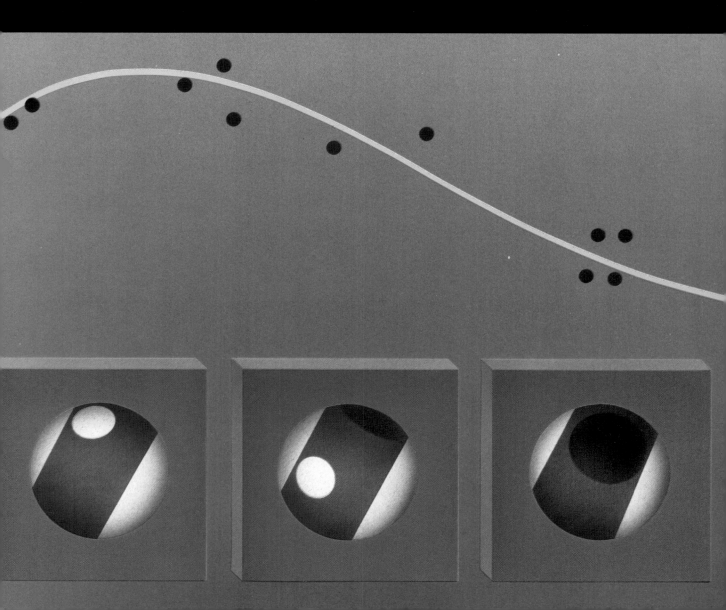

Clues From Eclipses

Pluto and Charon are engaged in a cosmic dance that gives scientists a rare chance to discern additional details of the two bodies. Charon's orbit is not ordinarily seen edge on. But in January 1985, the planet and its moon moved into an alignment that would cause them to eclipse each other repeatedly for nearly six years. Such eclipses can be observed from Earth only twice during the planet's 248-year circuit of the Sun; the next time will be in the twenty-second century, when Pluto's orbital position and the sixty-two-degree tilt of Charon's orbital plane again combine to carry the moon across Pluto's face. The satellite passes in front of the planet or behind it in a few hours, repeating the cycle every 6.4 days.

Eclipses help astronomers calculate the amount of light reflected by each member of the pair. When Charon disappears behind Pluto, scientists know that all the radiation they record is reflected by the planet alone. When Charon is not eclipsed, they subtract the known quantity for Pluto from the recorded light and identify the portion that is ordinarily reflected by the satellite. Eclipses have revealed that Charon's surface is a neutral gray, darker than Pluto's; its material seems to be water ice and perhaps rock, with little or no methane ice. Continuing studies of the duration and amount of dimming as different parts of the planet are obscured will let astronomers map the light and dark regions of Pluto and may provide more precise diameters for both bodies. With more accurate dimensions, scientists can refine figures for the mass and density of the system, which in turn may unlock further secrets of the worlds at the edge of the Solar System.

are the only bodies in the Solar System whose gravities have locked them in mutually synchronous orbit: They always show each other the same face, rigidly circling one another at a distance of about 12,000 miles.

To many astronomers, the odd couple represents not the end product of the Solar System's formation so much as a point near the beginning, a surviving remnant of the original stellar debris from which the other planets finally coalesced. If that is true, this outrider among known planets would be an illuminating archive of Solar System memorabilia, if one could only study it. But there is no immediate prospect for a probe. A spacecraft launched today, without the benefit of the alignment that permitted Voyager to draw energy from planetary gravities, would take about forty years to reach its goal. Not even *Voyager 2* could have made the trip. The trajectory to Pluto would have taken the probe through the very core of Neptune, an impossible option. Nevertheless, Pluto and its moon have recently become the subjects of intensive scientific study, revealing themselves as they have never done before.

The renewed interest in the pair stems from the fact that their orbit, from an Earth-based point of view, has brought them closer and closer together, so that in each week-long rotation they seem to merge, Pluto occulting Charon, then Charon passing in front of Pluto *(pages 124-125)*. The rhythmic rise and fall of light from the spinning duo yields spectral data that permits scientists to distinguish between the planet and its moon, and also to make their best— and lowest—estimates yet of the dimensions of the distant bodies. Once thought to have ten times the mass of the Earth, the Pluto-Charon pair is now believed to be just two-thousandths of an Earth mass.

With these observations and the help of supercomputers, astronomers have

Traveling in tandem, Pluto *(red)* and Charon *(gray)* trace complementary corkscrew paths through the outskirts of the Solar System. Since the tiniest planet is only about eight times more massive than its moon, each strongly affects the other's motion. (Other moons also influence their worlds, but enormous mass differences make such effects nearly imperceptible.) Pluto and Charon rapidly revolve around a shared center of gravity *(yellow)* located closer to Pluto; circling every 6.4 days, the pair completes 14,000 loops in one 248-year solar orbit.

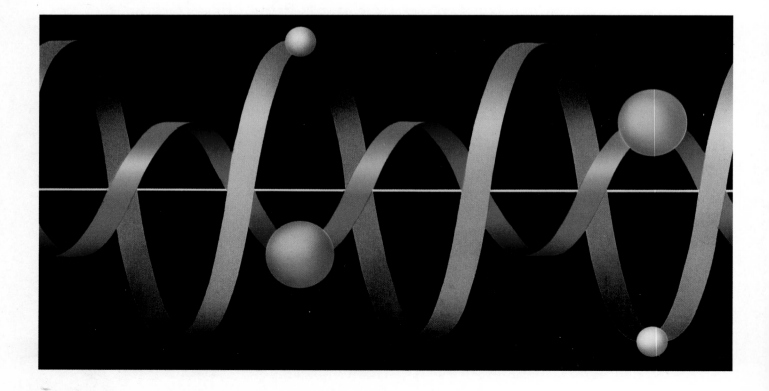

begun to model how the double planet actually looks and behaves. Pluto appears to be slightly redder than the dark, neutral face Charon holds toward it. Infrared data shows large polar ice caps of frozen methane extending almost halfway to the Plutonian equator, separated by a warmer, but still very chilly, equatorial band. Pluto's thin atmosphere—the equivalent of an Earth atmosphere only thirty feet thick, according to some estimates—appears to be rich in methane. Charon, in contrast, appears in infrared readings to be a world covered with water ice but devoid of methane in any form. If Charon originally possessed a surface store of methane, the gas may have long since sublimated from ice to vapor and escaped the moon's feeble gravity. Taking the lower set of dimensions offered thus far, astronomers calculate the average density of Pluto and Charon at about two grams per cubic centimeter, suggesting they are not mere chunks of water ice but may each possess a planetlike rocky core.

As the occulting dance continues, the system will be measured more precisely and many of its secrets made to unfold. But the central mystery, that of its origin, may be impenetrable at this distance. If Pluto really is an escaped moon of Neptune, how did it come to have a gargantuan satellite of its own? If Pluto and Charon are truly primordial objects left over from the formation of the Solar System, why are they not orbiting closer to the plane of the other planets? The answers could hide in the darkness beyond Pluto, perhaps in the form of an undiscovered body whose passage created the myriad contradictions of the outermost planets and their moons. Upon its discovery in 1930, Pluto was immediately hailed as the long-sought "Planet X," the object whose gravitational effect would account for differences between the observed and predicted positions of Uranus and Neptune. Some astronomers, however, later came to the conclusion that the newly found world lacked sufficient mass for its gravity to have such influence on its giant neighbors.

Somewhere out there, a massive tenth planet or a dim companion star to the Sun may be awaiting discovery. The gravitational pull of such an object would do more than perturb the outer planets' orbits. Now and then it would drag frozen comets out of their remote solar orbits and send them careering through the Solar System. But if this so-called Nemesis star or Planet X exists, it is well hidden: Its gravitational effects, if real, are too small to be measured precisely with existing instruments, and its presumptive remoteness from the Sun ensures that it is too faint to be detected among the teeming stars and galaxies crowding a telescope's field of view.

There is essentially no possibility that the probes heading outside the orbit of Pluto—the two *Pioneers,* the two *Voyagers*—will come upon this object by chance. Still, the little craft have reconnoitered the outer reaches of the Solar System better than anyone expected. Now a kind of robotic rest awaits them. For some years to come, they will coast through the void, murmuring quietly to anyone still listening on their native planet near the distant yellow Sun. Then, early in the next century, their radio voices will fail, and wrapped in silence, they will lose themselves in the encircling wilderness of stars.

Defenseless against meteors that raked through its wispy atmosphere, Neptune's moon Triton holds up a cratered face to the cloudy blue visage of its parent planet.

Rising in perpetual dusk above the frozen landscape of its companion moon Charon, Pluto reflects the paltry rays of a bright star *(center)*—all that these distant worlds ever see of the Sun, three and a half billion miles away.

APPENDIX

DENIZENS OF THE SOLAR SYSTEM

Pluto

Mercury

Venus

Saturn Mars Neptune Jupiter Uranus Earth

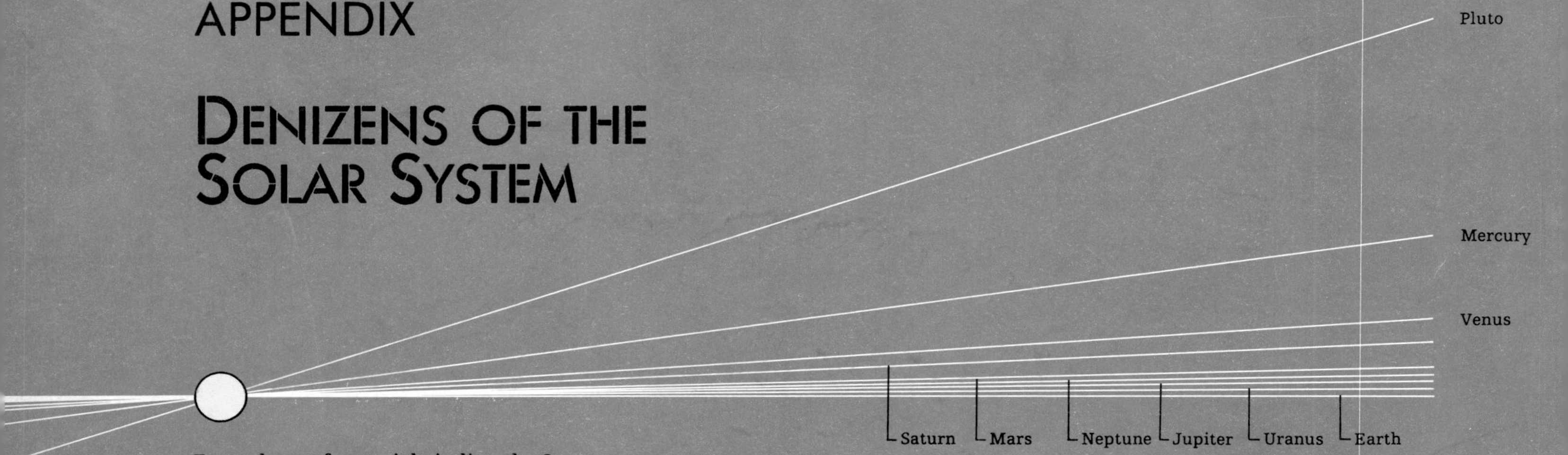

Formed out of material circling the Sun, the planets ultimately settled into orbits slightly angled to the plane of the ecliptic, the apparent path of the Sun across the sky. As seen above, the orbits of Mercury and Pluto, which are the most elliptical in the Solar System, are also the most inclined.

Although all of them are ruled by the same physical laws, there is considerable difference among the Sun's planets. For instance, they orbit at varying distances from the Sun, in paths shaped by the balance between solar gravity and their own motion. The way they are tipped with respect to their orbits largely defines their seasons, and their individual rates of spin set the length of each planetary day.

Scientists do not yet know the underlying causes of some planetary phenomena—why Uranus wallows around its orbit on its side, for example, or why Pluto's orbit differs so radically from those of its fellows—but much has been learned about these bodies over the years. On these two pages are diagrams that depict the planets' sizes, orientations, and orbits and a table *(right)* that highlights the salient traits of the denizens of the Solar System.

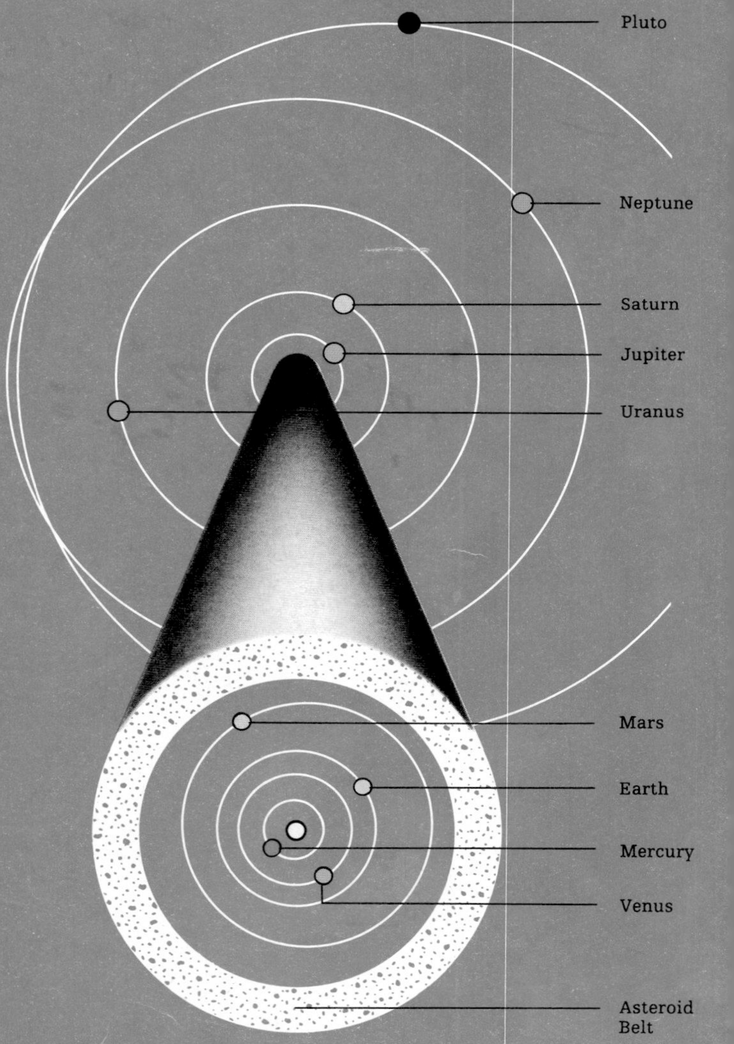

From a vantage above the north pole of the Sun, the orbits of most of the planets appear to be nearly circular. Pluto's orbit, in contrast, is so elliptical that the planet swings inside the orbit of Neptune for about 20 of every 248 years.

Pluto

Neptune

Saturn

Jupiter

Uranus

Mars

Earth

Mercury

Venus

Asteroid Belt

Planetary Data	Mercury	Venus	Earth	Mars	Jupiter	Saturn	Uranus	Neptune	Pluto
Equatorial Radius (Miles)	1,516	3,759	3,963	2,112	44,679	37,284	16,247	15,380	752
Mass (Trillion Trillion Pounds)	0.729	10.738	13.177	1.416	4,187.0	1,253.8	190.95	227.1	0.026
Mean Density (Earth = 1)	0.98	0.95	1.0	0.71	0.24	0.125	0.216	0.30	0.36
Gravity (Earth = 1)	0.39	0.88	1.0	0.38	2.34	0.93	0.79	1.13	0.0637
Period of Rotation (Hours)	1,407.6	5,832.2	23.9	24.6	9.8	10.2	15.5	15.8	6.4
Escape Velocity (Miles per Hour)	9,619	23,042	25,055	11,185	133,104	79,639	47,470	52,794	2,640
Major Atmospheric Gas	Oxygen	Carbon Dioxide	Nitrogen	Carbon Dioxide	Hydrogen	Hydrogen	Hydrogen	Hydrogen	Methane
Inclination of Equator (Degrees)	0.0	2.6	23.5	25.2	3.1	26.7	82.1	29.0	62.0
Known Moons	0	0	1	2	16	17	15	2	1
Eccentricity of Orbit	0.206	0.007	0.017	0.093	0.048	0.056	0.047	0.009	0.246
Mean Orbital Velocity (Miles per Hour)	107,132	78,364	66,641	53,980	29,216	21,565	15,234	12,147	10,604
Minimum Distance from Sun (Millions of Miles)	28.6	66.8	91.4	128.4	460.3	837.6	1,699.0	2,771.0	2,756.0
Maximum Distance from Sun (Millions of Miles)	43.4	67.7	94.5	154.9	507.2	936.2	1,868.0	2,819.0	4,555.0
Mean Distance from Sun (Millions of Miles)	36.0	67.2	93.0	141.6	483.4	886.7	1,784.0	2,794.4	3,656.0
Period of Revolution (Earth Years)	0.24	0.62	1	1.88	11.86	29.46	84.01	164.79	247.70
Inclination of Orbit to Plane of Ecliptic (Degrees)	7.00	3.39	—	1.85	1.31	2.49	0.77	1.77	17.15

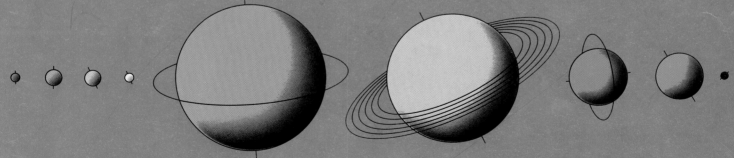

All of the planets except Mercury rotate around an axis tilted with respect to the plane of their orbit (above). Mercury's axis is not inclined, while Uranus and Pluto display the most extreme inclinations, spinning almost on their sides.

Diagramed below, the planets' average relative distances from the Sun stretch more than three billion miles from the asteroid belt to the distant track of Pluto.

GLOSSARY

Absorption line: a dark line or band at a particular wavelength on a spectrum, formed when a substance between a light source and an observer absorbs light of that wavelength. Different substances produce characteristic patterns of absorption lines.

Acceleration: a change in velocity.

Albedo: the reflectivity of a body; in planetary astronomy, the fraction of sunlight reflected by a planet or moon.

Angstrom: a unit of length equal to one ten-billionth of a meter (about four billionths of an inch); used in astronomy as a measure of wavelengths.

Angular diameter: an object's width on the sky, measured in units of arc distance—degrees, minutes (sixty per degree), and seconds (sixty per minute). The Moon's angular diameter is just over half a degree; Jupiter's is forty-seven seconds of arc.

Angular momentum: a measure of an object's inertia, or state of motion, about an axis of rotation.

Antenna: a device that beams and collects radio waves; in radio astronomy, the component of a radio telescope that converts received radio waves into electrical signals.

Asteroid: a small, rocky, star-orbiting body with no atmosphere.

Astronomical Unit (AU): a distance unit, often used within the Solar System, that is equal to the mean distance between the Earth and the Sun, or 92,960,116 miles.

Atmosphere: a gaseous shell surrounding a planet or other body. Also, a unit of pressure equivalent to the air pressure at sea level on Earth, about 14.7 pounds per square inch.

Atom: the smallest component of a chemical element that retains the properties associated with that element. Atoms are composed of protons, neutrons, and electrons.

Attitude: the orientation of an object with respect to its direction of motion.

Aurora: light given off by collisions between charged particles trapped in a planet's magnetic field and atoms of atmospheric gases near a planet's magnetic poles.

Bending wave: in a planetary ring, a corrugated wave pattern caused by the gravitational pull of a moon alternately orbiting above and below the ring plane.

Bit: the smallest unit of computerized information, represented by a single zero or one (the term is an abbreviation of "binary digit"). Streams of bits are used to transmit images and other data between unmanned spacecraft and controllers on Earth. *See* byte.

Bow shock: in planetary science, the boundary region where the solar wind is first deflected by a planet's magnetic field.

Brightness: the amount of light received from an object; a combined result of its actual luminosity, its distance, and any light absorption by intervening dust or gas.

Byte: a common unit of computerized information, usually consisting of eight bits. *See* bit.

Caldera: a crater formed by the collapse or subsidence of the central part of a volcano.

Cassini Division (also Cassini's Division): the seeming gap, now known to include several narrow rings, between Saturn's two brightest rings; named for the Italian astronomer Giovanni Cassini, who discovered it in 1675.

Celsius: a scientific temperature scale in which 0 degrees is the freezing point and 100 degrees the boiling point of water at one atmosphere.

Comet: an asteroid-size body of dusty ice that travels an elongated orbit around the Sun.

Continuous spectrum: a spectrum consisting of all wavelengths in a given range, without absorption or emission lines.

Convection: the process of heat transfer caused by thermal variations in a fluid.

Core: the central component of a celestial body. In planets it is usually composed of dense, hot material, frequently solid.

Coriolis effect: the apparent deflection of an object's trajectory over the surface of a rotating body—a consequence of the body's rotation. The effect is seen, for example, in the spiral shapes taken by storms on Earth and in the atmospheres of other planets.

Cosmic rays: atomic nuclei or other charged particles moving at close to the speed of light.

Data compression: any of several methods used to reduce the number of bits required to transmit computerized information in the least amount of time.

Density wave: a moving pattern of compression and rarefaction. Density waves are thought to produce the grooved structure of Saturn's rings.

Differential rotation: rotation in which components at different distances from the center orbit at different rates rather than as a rigid body.

Doppler effect: a wave phenomenon in which waves are compressed as their source approaches the observer, thus increasing their frequency, or are stretched out as the source recedes from the observer, decreasing the frequency.

Dust: an interplanetary or interstellar material consisting of individual, widely separated, microscopic particles of rock, often coated with ice.

Eclipse: the obscuration of light from a celestial body as it passes through the shadow of another body. In a *lunar eclipse,* the Earth's shadow darkens the Moon as the Earth passes between the Sun and the Moon. In a *solar eclipse,* the Moon passes between the Sun and the Earth, blocking, or occulting, the Sun. *See* occultation.

Electromagnetic radiation: radiation consisting of periodically varying electric and magnetic fields that vibrate perpendicularly to each other and travel through space at the speed of light.

Electromagnetic spectrum: the array, in order of frequency or wavelength, of electromagnetic radiation, from low-frequency, long-wavelength radio waves to high-frequency, short-wavelength gamma rays.

Electron: a negatively charged particle that normally orbits an atom's nucleus but may exist independently.

Ellipse: a circle in which the vertical and horizontal axes are not equal in length, resulting in a symmetrical oval. The orbits of all Solar System planets are ellipses.

Emission line: a bright line or band at a particular wavelength on a spectrum, emitted directly by the source and indicating a chemical constituent of that source.

Epicycle: a small circle centered on the rim of a larger one, used by early astronomers to explain the motions of the planets in what was thought to be an Earth-centered universe.

Equator: the imaginary line around a celestial object that lies in a plane passing through the center of the object perpendicular to its spin axis.

Escape velocity: the speed and direction of motion that are required for an object to move beyond a closed orbit around a planet or other body.

Fahrenheit: a nonastronomical temperature scale in which 32 degrees is equivalent to 0 degrees Celsius and 212 degrees is equivalent to 100 degrees Celsius.

Frequency: for electromagnetic waves, the number of oscillations past a point per unit of time, expressed in hertz (cycles per second). *See* wavelength.

Gamma rays: the most energetic form of electromagnetic radiation, with the highest frequency and shortest wavelength.

Gas: a fluid state of matter in which atoms and molecules are not bound together and will diffuse unless contained.

Gravity: the mutual attraction of separate masses; a fundamental force of nature.

Guide star: a star used by spacecraft navigation systems as a course marker; also a star tracked by a guide telescope during a long photographic exposure to ensure that the main telescope is accurately aligned.

Helium: the second-lightest chemical element and the second most abundant in the universe.

Hydrocarbon: a chemical compound made up of carbon and hydrogen atoms. Because carbon atoms readily form strong bonds, thousands of types of hydrocarbons are possible; seven have been found in interstellar material.

Hydrogen: the most common element in the universe and the chief constituent of many outer planets. *See* liquid metallic hydrogen.

Hyperbola: a type of open loop that is sometimes followed by bodies with too high a velocity to travel a closed orbit.

Ice: the frozen crystalline state of a substance.

Imaging: the techniques used to construct images from digital information produced by electronic sensors.

Image enhancement: the numerical manipulation of digital image data to bring out otherwise unnoticeable detail.

Infrared: a band of electromagnetic radiation with a lower frequency and a longer wavelength than visible red light.

Intensity: the amount of radiation received from an object. Optical astronomers prefer the term *brightness*.

Interferometer: in astronomy, an arrangement of two or a few telescopes used in tandem to locate and examine sources of electromagnetic emission. With components separated by a distance several times the wavelength of incoming waves, an interferometer takes advantage of the natural interference of electromagnetic waves as they arrive from different directions.

Inverse-square law: the mathematical description of how some forces, including electromagnetism and gravity, weaken in inverse proportion to the square of the distance from the source. For example, the strength of a planet's gravitational field at two million miles is four times less than it is at one million miles.

Ion: an atom that has lost or gained one or more electrons and has become electrically charged. In comparison, a neutral (uncharged) atom has an equal number of electrons and protons, giving the atom a zero net electrical charge. A positive ion has fewer electrons than the neutral atom; a negative ion has more.

Jet stream: a high-speed, high-altitude wind current blowing generally in the direction of a planet's rotation.

Kelvin: an absolute temperature scale that uses Celsius degrees but sets 0 at absolute zero, about minus 273 degrees Celsius.

Kepler's laws: three laws of planetary motion formulated by Johannes Kepler in the 1600s: that planets follow elliptical orbits, that the line between the Sun and a planet sweeps through equal areas in equal times (so that the closer a planet is to the Sun, the faster it moves), and that a planet's orbital period is directly related to its average distance from the Sun.

Kinetic energy: an object's energy of motion; for example, the force of a falling body. *See* potential energy.

Latitude: on a celestial body, the angular distance north or south of the equator. *See* equator.

Light: the visible part of the electromagnetic spectrum.

Liquid metallic hydrogen: an exotic, highly compressed form of liquid hydrogen that readily conducts electricity.

Longitude: on a celestial body, the angular distance east or west of a standard meridian, or great circle, passing through the body's rotational poles.

Luminosity: an object's total energy output, usually measured in ergs per second.

Magnetic field lines: lines of magnetic force extending from one magnetic pole to the other. The space traversed by these lines is the magnetic field.

Magnetic pole: one of two locations on a planet's surface where its magnetic field lines converge, as opposed to the geographic poles defined by a planet's axis of rotation.

Magnetometer: a device for measuring the strength and direction of a magnetic field.

Magnetopause: the well-defined boundary between a planet's magnetosheath and its magnetosphere.

Magnetosheath: the turbulent region between the bow shock and the magnetopause where solar wind particles are slowed and deflected around a planet's magnetosphere.

Magnetosphere: the region around a planet in which its magnetic field is the dominant magnetic influence; the magnetosphere is bounded by the magnetopause.

Mass: a measure of the total amount of material in an object, determined either by the object's gravity or by its tendency to stay in motion, if in motion, or at rest, if at rest.

Molecule: the smallest component of a chemical that retains that chemical's properties. A molecule may consist of one or more atoms bonded together.

Moon: one of a planet's natural satellites, generally no smaller than ten miles in diameter. There are more than fifty known moons in the Solar System, including Earth's.

Neutron: an uncharged particle with a mass similar to a proton's; normally found in an atom's nucleus.

Noise: meaningless, random changes in radiation that tend to obscure a specific signal.

Nonthermal emission: an electromagnetic radiation pattern that increases in intensity as frequency increases, as, for example, in synchrotron radiation. *See* thermal emission.

Nuclear fusion: the combining of two atomic nuclei to form a heavier nucleus, a process that releases a great deal of energy.

Nucleus: the center of an atom, composed of protons and neutrons and orbited by electrons.

Occultation: an event in which one celestial body passes in front of another, partially or totally obscuring it.

Orbit: the path of an object revolving around another object or point.

Particle: a fundamental component of matter, such as a molecule, atom, proton, neutron, or electron; also, the small fragments of material in planetary rings.

Penumbra: the area of lighter shadow surrounding the central shadow cast by an illuminated body. *See* umbra.

Photometer: a device for measuring the intensity of light and ultraviolet and infrared radiation.

Photon: a unit of electromagnetic energy associated with a specific wavelength.

Photopolarimeter: a specialized telescope that measures changes in the amount of light coming from a source, enabling scientists to infer density, structure, and other properties of the source.

Pixel (also picture element): in a digital image, a single dot numerically assigned an intensity that can be represented as real or false color.

Planet: a large, nonstellar body that orbits a star and shines only with reflected light.

Plasma: a gaslike association of ionized particles that responds collectively to electric and magnetic fields. Because plasma particles do not interact the way particles of ordinary gas do, plasma is considered a fourth state of matter, along with solid, liquid, and gas.

Pole: one of two points on a planet's surface where its rotational axis ends; a planet's magnetic poles may or may not coincide with those formed by the rotational axis.

Potential energy: an object's stored energy; for example, the latent force of a body poised to fall. *See* kinetic energy.

Probe: an automated, crewless vehicle used in space exploration.

Proton: a positively charged particle with about 2,000 times the mass of an electron; normally found in an atom's nucleus.

Radiation belt: a band of charged particles trapped in a planet's magnetic field; for example, the Earth's Van Allen belts.

Radio: the least energetic form of electromagnetic radiation, with the lowest frequency and the longest wavelength.

Radio astronomy: the observation and study of radio waves produced by astronomical objects.

Radio burst: an abrupt, strong increase in radio emissions from an astronomical object.

Radioisotope thermoelectric generator (RTG): a type of nuclear power supply, carried on many deep-space probes, in which electricity is generated by the heat released in the decay of a radioactive isotope.

Radiometer: any device that measures the intensity of radiation.

Radio telescope: an instrument for studying astronomical objects at radio wavelengths.

Receiver: in radio astronomy, a device for detecting and measuring radio waves collected by an antenna.

Resolution: the degree to which details in an image can be separated, or resolved. The resolving power of a telescope is usually proportional to the diameter of its collector.

Resonance: in astronomy, the enhanced gravitational effect on a smaller body when its orbital period is an exact fraction of a larger neighbor's.

Retrograde motion: real or apparent motion against the prevailing direction of movement. The apparent backward motion of planets as they are overtaken by the Earth is an illusion of retrograde motion; Triton's orbit against the direction of Neptune's rotation is real.

Roche limit: the boundary within which a planet's tidal forces both prevent the formation of moons and destroy orbiting moons that stray inside the limit; named after its discoverer, French astronomer Edouard Roche.

Satellite: any body, natural or artificial, in orbit around a planet; used most often to describe moons and spacecraft.

Scan platform: on a space probe, the movable component holding cameras and other scientific equipment that needs to be aimed.

Shepherd moons: small moons, sometimes paired, that gravitationally influence the orbits of particles in some planetary rings.

Shock wave: in astronomy, a sudden discontinuity in the flow of a gas, liquid, or plasma characterized by abrupt increases in temperature, pressure, and velocity.

Solar system: the Sun and its associated family of planets, asteroids, and other orbiting bodies; more generally, any star and the objects that orbit it.

Solar wind: a current of charged particles that streams outward from the Sun.

Speckle interferometry: a technique for obtaining a high-resolution image of a planet or star by using a computer to analyze the speckled patterns of light produced by atmospheric distortion of multiple high-speed photographs of the body.

Spectrogram: a photographic image of an astronomical spectrum.

Spectrograph: an instrument that splits light or other electromagnetic radiation into its individual wavelengths, or spectrum, and records the result photographically.

Spectroscopy: the study of spectra, including the position and intensity of emission and absorption lines, to learn about the physical processes that create them.

Spectrum: the array of colors or wavelengths obtained by dispersing light, as through a prism; often banded with absorption or emission lines, which can be interpreted to show the chemistry and motion of the light source.

Spin axis: the line around which an object rotates; in planets and stars, it extends through the center of the planet between the north and south geographic poles.

Synchronous rotation: a phenomenon in which a moon spins on its axis exactly once per orbit, thus keeping the same face toward its planet at all times; seen in most moons in the Solar System.

Synchrotron radiation: a type of nonthermal emission generated by electrons and other charged particles spiraling around magnetic field lines at near light-speed.

Thermal emission: an electromagnetic radiation pattern that decreases in intensity as frequency increases; produced by heat-related processes. *See* nonthermal emission.

Tidal force: the force generated in a body by variations in the gravitational attraction exerted by another body.

Trajectory: the curved path of a moving body.

Ultraviolet: a band of electromagnetic radiation with a higher frequency and shorter wavelength than visible blue light.

Umbra: the dark central shadow cast by an illuminated body.

Universe: the sum total of all matter and radiation and the space they occupy.

Velocity: the speed and direction of motion.

Wave: the propagation of a pattern of disturbance.

Wavelength: the distance from crest to crest of a wave.

X-rays: a band of electromagnetic radiation intermediate in wavelength between ultraviolet radiation and gamma rays. Because x-rays are absorbed by the atmosphere, x-ray astronomy is performed in space.

BIBLIOGRAPHY

Books

Abell, George O., David Morrison, and Sidney C. Wolff, *Exploration of the Universe.* New York: Saunders College Publishing, 1987.

Audouze, J., and G. Israel, eds., *Cambridge Atlas of Astronomy.* New York: Cambridge University Press, 1985.

Beatty, J. Kelly, Brian O'Leary, and Andrew Chaikin, eds., *The New Solar System.* Cambridge, Mass.: Sky, 1981.

Blanchard, Paul A., *Atoms and Astronomy.* Washington, D.C.: NASA, 1976.

Briggs, Geoffrey, and Fredric Taylor, *The Cambridge Photographic Atlas of the Planets.* New York: Cambridge University Press, 1986.

Clarke, Arthur C., *The Promise of Space.* New York: Harper & Row, 1968.

Elliot, James, and Richard Kerr, *Rings.* Cambridge, Mass.: MIT Press, 1984.

Ferris, Timothy, *SpaceShots.* New York: Pantheon, 1984.

Frazier, Kendrick, and the Editors of Time-Life Books, *Solar System* (Planet Earth series). Alexandria, Va.: Time-Life Books, 1985.

Gallant, Roy A., *Picture Atlas of Our Universe.* Washington, D.C.: National Geographic Society, 1980.

Gehrels, Tom, and Mildred Shapley Matthews, eds., *Saturn.* Tucson: University of Arizona Press, 1984.

Greenberg, Richard, and André Brahic, *Planetary Rings.* Tucson: University of Arizona Press, 1984.

Hartmann, William K., R. J. Phillips, and G. J. Tay, eds., *Origin of the Moon.* Houston, Tex.: Lunar & Planetary Institute, 1986.

Henbest, Nigel, and Michael Marten, *The New Astronomy.* Cambridge: Cambridge University Press, 1983.

Hoskin, M. A., ed., *Journal for the History of Astronomy.* Vol. 12. Chalfont St. Giles, England: Science History Publications Ltd., 1981.

Hunt, Garry, and Patrick Moore:
Jupiter. New York: Rand McNally, 1981.
Saturn. New York: Rand McNally, 1982.

Jones, Barrie William, *The Solar System.* Elmsford, N.Y.: Pergamon Press, 1984.

Kaufmann, William J., III, *Universe* (2nd ed.). New York: W. H. Freeman, 1987.

Kivelson, Margaret G., ed., *The Solar System: Observations and Interpretations.* Rubey Vol. 4. Englewood Cliffs, N.J.: Prentice-Hall, 1986.

Koestler, Arthur, *The Sleepwalkers.* London: Hutchinson, 1968.

La Cotardière, Philippe de, ed., *Larousse Astronomy.* New York: Facts on File, 1987.

Learner, Richard, *Astronomy through the Telescope.* New York: Van Nostrand Reinhold, 1981.

Lewis, John S., and Ronald G. Prinn, *Planets and Their Atmospheres: Origin and Evolution.* Orlando, Fla.: Academic Press, 1984.

Miller, Ron, and William K. Hartmann, *The Grand Tour: A Traveler's Guide to the Solar System.* New York: Workman, 1981.

Mitton, Simon, ed., *The Cambridge Encyclopaedia of Astronomy.* New York: Crown, 1977.

Moore, Patrick, ed., *The International Encyclopedia of Astronomy.* New York: Orion Books, 1987.

Morrison, David, *Voyages to Saturn.* Washington, D.C.: NASA, 1982.

Morrison, David, ed., *Satellites of Jupiter.* Tucson: University of Arizona Press, 1982.

Morrison, David, and Jane Samz, *Voyage to Jupiter.* Washington, D.C.: NASA, 1980.

Murray, Bruce, and Carl Sagan, *The Planets.* New York: W. H. Freeman, 1983.

Nicolson, Iain, *Gravity, Black Holes and the Universe.* New York: Halsted Press, 1981.

Sidgwick, John B., *William Herschel: Explorer of the Heavens.* London: Faber and Faber, 1953.

Snow, Theodore P., *Essentials of the Dynamic Universe.* St. Paul: West, 1987.

Space, by the Editors of Time-Life Books (Understanding Computers series). Alexandria, Va.: Time-Life Books, 1987.

Suess, Hans E., *Chemistry of the Solar System.* New York: John Wiley & Sons, 1987.

Washburn, Mark, *Distant Encounters.* San Diego: Harcourt Brace Jovanovich, 1983.

Whipple, Fred L., *Orbiting the Sun.* Cambridge, Mass.: Harvard University Press, 1981.

Yeates, C. M., et al., *Galileo: Exploration of Jupiter's System.* Washington, D.C.: NASA, 1985.

Zeilik, Michael, and John Gaustad, *Astronomy: The Cosmic Perspective.* New York: Harper & Row, 1983.

Zeilik, Michael, and Elske v. P. Smith, *Introductory Astronomy and Astrophysics.* Philadelphia: Saunders College Publishing, 1987.

Periodicals

Anderson, Ian, "The Pock Marked Face of Miranda." *New Scientist,* February 6, 1986.

Augensen, Harry, and Jonathan Woodbury, "The Electromagnetic Spectrum." *Astronomy,* June 1982.

"Barely a Planet: New Images of Pluto." *New York Times,* September 15, 1987.

Beatty, J. Kelly:
"Pluto and Charon: The Dance Begins." *Sky & Telescope,* June 1985.
"Pluto and Charon: The Dance Goes On." *Sky & Telescope,* September 1987.

Bennett, Gary L., "Return to Jupiter." *Astronomy,* January 1987.

Berry, Richard:
"Uranus: The Voyage Continues." *Astronomy,* April 1986.
"Voyager: Discovery at Uranus." *Astronomy,* May 1986.

Brown, Robert Hamilton, and Dale P. Cruikshank, "The Moons of Uranus, Neptune and Pluto." *Scientific American,* July 1985.

Chaikin, Andrew, "Voyager among the Ice Worlds." *Sky & Telescope,* April 1986.

Croswell, Ken:
"A Mission to Pluto." *Space World,* September 1987.
"Pluto: Enigma on the Edge of the Solar System." *Astronomy,* July 1986.

Cuzzi, Jeffrey N., "Ringed Planets: Still Mysterious—II." *Sky & Telescope,* January 1985.

Cuzzi, Jeffrey N., and Larry W. Esposito, "The Rings of Uranus." *Scientific American,* July 1987.

Darling, David J.:
"Spectral Visions, Part 1: The Long Wavelengths." *Astronomy,* August 1984.

"Spectral Visions, Part 2: The Short Wavelengths and Beyond." *Astronomy,* September 1984.

Daughtrey, T., "Warming Up to a Cold and Distant World." *Space World,* January 1985.

Dickerson, Richard E., "Chemical Evolution and the Origin of Life." *Scientific American,* September 1978.

Eberhart, Jonathan:
"Pluto: Limits on Its Atmosphere, Ice on Its Moon." *Science News,* September 26, 1987.
"Voyager: Setting Safe Sights for Neptune." *Science News,* November 29, 1986.

"Eclipse of Pluto Observed." *USA Today,* June 1985.

"ESA Defines Saturn Moon Probe for Proposed NASA Mission in 1990s." *Aviation Week & Space Technology,* April 1, 1985.

Esposito, Larry W., "The Changing Shape of Planetary Rings." *Astronomy,* September 1987.

Flasar, F. Michael, "Global Dynamics and Thermal Structure of Jupiter's Atmosphere." *Icarus,* 1986.

Franklin, Kenneth L.:
"Gravitational Forces and Effects." *Natural History,* October 1963.
"Gravitational Forces and Effects: Part II." *Natural History,* November 1963.
"Radio Waves from Jupiter." *Scientific American,* July 1964.

"Frigid Oceans for Triton and Titan." *Science,* July 29, 1983.

"Getting Small." *Scientific American,* April 1987.

Gold, Michael, "Voyager to the Seventh Planet." *Science,* May 1986.

Gore, Rick, "Uranus: Voyager Visits a Dark Planet." *National Geographic,* August 1986.

Graham, John, "The Voyager 2 Team Reflects on a 'Wonderful Mission.' " *Astronomy,* May 1986.

Gupta, S., E. Ochiai, and C. Ponnamperuma, "Organic Synthesis in the Atmosphere of Titan." *Nature,* October 29, 1981.

Harrington, R. S., and B. J. Harrington, "Pluto: Still an Enigma after 50 Years." *Sky & Telescope,* June 1980.

Henbest, Nigel, "Uranus after Voyager." *New Scientist,* July 31, 1986.

Hubbard, W. B., "Observations of Uranus Occultation Events." *Nature,* July 7, 1977.

Ingersoll, Andrew P., "Uranus." *Scientific American,* January 1987.

Johnson, Torrence V., Robert Hamilton Brown, and Laurence A. Soderblom, "The Moons of Uranus." *Scientific American,* April 1987.

Kieffer, Susan Werner, "Volcanoes and Atmospheres: Catastrophic Influences on the Planets." *Earthquakes & Volcanoes,* 1986.

Kohlhase, Charles E., "Aiming at Neptune." *Astronomy,* November 1987.

Laeser, Richard P., William I. McLaughlin, and Donna M. Wolff, "Engineering Voyager 2's Encounter with Uranus." *Scientific American,* November 1986.

Lunine, J. I., David J. Stevenson, and Yuk L. Yung, "Ethane Ocean on Titan." *Science,* December 16, 1983.

Marbach, William D., and Michael D. Cantor, "The Search for Planet X." *Newsweek,* July 13, 1987.

"Mission Status Report No. 79: The Magnetosphere." *Voyager Bulletin,* February 12, 1986.

Morabito, L. A., et al., "Discovery of Currently Active Extraterrestrial Volcanism." *Science,* June 1, 1979.

Morrison, N. D., and S. Gregory, "The Exotic Atmosphere of Titan." *Mercury,* September/October 1985.

"Neptune's Moon Has Atmosphere!" *New Scientist,* October 20, 1983.

Northrop, T. G., A. G. Opp, and J. H. Wolfe, "Pioneer 11 Saturn Encounter." *Journal of Geophysical Research,* November 1, 1980.

Overbye, Dennis, "Voyager Was on Target Again." *Discover,* April 1986.

Owen, Tobias, "Time Travel and Chemical Evolution." *Planetary Report,* November/December 1987.

Sackett, P. D., "Titan Air Resembles That of Prebiotic Earth." *Science News,* June 25, 1983.

Sagan, Carl, "Voyager's Triumph." *Popular Science,* October 1986.

Schultz, Ron, "Interview with Bradford Smith." *Omni,* February 1987.

Soderblom, Laurence A., "The Galilean Moons of Jupiter." *Scientific American,* January 1980.

Stone, E. C., and E. D. Miner, "The Voyager 2 Encounter with the Uranian System." *Science,* July 4, 1986.

Sykes, Mark V., et al., "IRAS Serendipitous Survey Observations of Pluto and Charon." *Science,* September 11, 1987.

"Titan, the Smog-Shrouded." *Astronomy,* December 1981.

Toon, Owen B., and Steve Olson, "The Warm Earth." *Science 85,* October 1985.

"Uranus is Perturbed: Usual Suspect Rounded Up." *Discover,* September 1987.

Van Allen, James, "Interplanetary Particles and Fields." *Scientific American,* September 1975.

West, Robert A., Darrell F. Strobel, and Martin G. Tomasko, "Clouds, Aerosols, and Photochemistry in the Jovian Atmosphere." *Icarus,* 1986.

Other Publications

Beynon, W. J. G., et al., eds., "Voyager Project." *Space Science Reviews,* November/December 1977.

Burbidge, Geoffrey, David Layzer, and John G. Phillips, eds., *Annual Review of Astronomy and Astrophysics.* Vol. 20. Palo Alto, Calif.: Annual Reviews, 1982.

"Cassini: Saturn Orbiter and Titan Probe." European Space Administration/NASA assessment study, August 1985.

"How We Get Pictures from Space." NASA Publication No. NF-151, no date.

Mitton, Simon, ed., *The Quarterly Journal of the Royal Astronomical Society.* Vol. 17. Oxford: Blackwell Scientific Publications, 1976.

"Pioneer Odyssey." NASA Publication No. SP-349.

"To Uranus and Beyond." NASA/Jet Propulsion Laboratory Publication No. JPL 400-303, September 1987.

"Voyager at Neptune and Triton: 1989." NASA/Jet Propulsion Laboratory Publication No. JPL 400-230, May 1984.

"Voyager at Uranus: 1986." NASA/Jet Propulsion Laboratory Publication No. JPL 400-268, July 1985.

"The Voyager Flights to Jupiter and Saturn." NASA Publication No. EP-191, no date.

"Voyager 1986." NASA press kit, January 1986.

"The Voyager Uranus Travel Guide." NASA/Jet Propulsion Laboratory Publication No. JPL D-2580, August 1985.

INDEX

Numerals in italics indicate an illustration of the subject mentioned.

Metallic hydrogen, liquid, *70,* 71
Meteoroid detector panel, *41*
Mimas (moon of Saturn), *96,* 103
Miranda (moon of Uranus), *6-7,*
    108-109, 116-*117*
Moons, 14-15; Galileo's observations
    of, *18-19;* of Jupiter, *2-3,* 18, 19,
    21, 49, 55, *61-64,* 65, *66-68;* naming
    of, 21; of Neptune, 21, 119-120,
    121, 130-131; occultation of, *28-29,*
    126-127; of Pluto, 121, *122, 123,*
    126-127, 128-129, *132-133;* of
    Saturn, *4-5,* 21, 82, *85,* 86, 92, *93,*
    94, 95, *96*-97, 103, *106-107;* of
    Uranus, *6-7,* 21, *108-109,* 116-*117*
Morabito, Linda, 65
Motion, planetary: of Pluto, 120-121,
    124-125, 128; Saturn's rotation,
    timing, 81. *See also* Orbits

N

Neptune (planet), 14-15, 118-119, *120,*
    130-131; discovery of, studies
    leading to, 27-32; moons of, 21,
    119-120, *121, 130-131*
Nereid (moon of Neptune), 119
Newton, Isaac, *18-19;* gravitational
    laws of, 18, 22, 24; Jupiter's mass
    estimated by, 47, 69; and solar
    spectrum, 49

O

Oberon (moon of Uranus), 116
Occultation: of moons, *28-29, 126-*
    127; in particle-beam astronomy,
    84; of stars, *26-27, 30-31,* 89, *119*
Orbits: earliest explanations of,
    15-16; gravity and, *24-25;* of
    Jupiter's moons, *19, 61, 68;*
    Kepler's laws of, 17-18; of Nep-
    tune's moons, 119; of Pluto and
    Charon, 120-121, 125, *126-127, 128;*
    retrograde motion in, 15, *16;* of
    Saturn's moons, 97, 103, *106;*
    Uranus's irregularities, study of,
    27-29, 30
Organic molecules, *95*
Outer Planets Grand Tour Project, 57

P

Pandora (moon of Saturn), 86,
    106-107
Particle-beam astronomy, 84
Particle detectors, *41*
Peale, Stanton, 65-66
Photopolarimeters, use of, 39; in
    Voyager 2's Saturn mission, 89
Pickering, William, 32, 34
Pioneer spacecraft, 37, 56-57; *Pioneer*
    10, 37, *39,* 56; *Pioneer 11* explora-
    tion of Saturn, 80-81, *83-*84
Planets: formation of, 47; gravita-
    tional fields of, *22-25;* naming of,

21, 26; place in Solar System,
    12-13, 14-*15*
Plasma (charged particles): in
    Saturn's B ring, theory of, 104;
    solar wind, 40, 54, *76*
Pluto (planet), 15, 21, *34,* 120-121,
    122, 123, *128,* 129; brightness
    studies of, *124-125,* 127; eclipse
    studies of, *126-127;* landscape of,
    132-133; moon of, 121, *122, 123,*
    126-127, 128-129, *132-133;* origin,
    theories of, *123;* search for, 32-36
Potential vs. kinetic energy, 24
Prometheus (moon of Saturn), 86,
    106-107
Ptolemy, Claudius, 16

R

Radiation belts, 54-55, 56
Radio astronomers and astronomy,
    48, 52-55, 115
Radios for Voyager probes, 58, 59-60;
    Uranus, transmissions from,
    112-113
Retrograde motion, 15, *16*
Reynolds, Ray, 66
Roche limit, 101

S

Saturn (planet), *4-5,* 14, 80-98;
    atmosphere of, 86, 89, 90, 91-92;
    composition of, 91; helium rain,
    theory of, 90-91; history of the
    study of, 81-82, *87;* magnetic field
    of, 92; moons of, *4-5,* 21, 82, *85,* 86,
    92, *93, 94,* 95, *96*-97, 103, *106-107;*
    Pioneer 11 exploration of, 80-81,
    83-84; rings of, *78-79,* 80, *83,* 84,
    85-86, *87, 88,* 89, 97-98, *99-107;*
    Voyager images of, *42-43, 85,* 86,
    88; Voyager missions to, 85-90
Shepherd moons, of Saturn's F ring,
    86, *106-107*
Slipher, Vesto, 34, 35, 46
Smith, Bradford, 65, 86; quoted, 60,
    86, 117, 118
Soderblom, Laurence, 116
Solar wind, *40,* 54, 76
Speckle interferometry, use of, *122*
Spectroscopy and spectrograms,
    46-47, *48-51*
Spectrum, electromagnetic, *33*
Spiral density waves, and Saturn's
    rings, 98, *102-103*
Spokes, in Saturn's rings, 85, 89,
    104-105
Stars, *120;* Herschel's observations
    of, 20; occultation of, *26-27, 30-31,*
    89, *119*
Sulfur and sulfur dioxide, Io's, *2-3,*
    61, *62-63*
Sunlight, spectrogram of, *48-49*
Synchrotron radiation, 54

T

Tatel, Howard, 52
Terrile, Rich, 85, 86
Tethys (moon of Saturn), *85,* 97
Thermodynamic law, 53
Titan (moon of Saturn), *4-5,* 82, 92,
    96; atmosphere of, 82, 92, 93, *94,*
    95, *96;* Cassini mission to,
    planned, *93*
Titania (moon of Uranus), 116
Tombaugh, Clyde, 34-36, *35*
Tori: around Io, 66, *68, 77;* around
    Saturnian complex, 96
Triton (moon of Neptune), 119, *121,*
    130-131

U

Umbriel (moon of Uranus), *108-109,*
    116, 117
Uranus (planet), *6-7,* 14; atmosphere
    of, *114*-115; Herschel's discovery
    of, 20-21; irregular behavior of,
    27-29, 30; magnetic field of, 115;
    moons of, *6-7,* 21, *108-109,* 116-
    117; naming of, 26; rings of, *30-31,*
    117-118, *118-119;* Voyager images
    of, *108-109, 114, 118-119;* Voyager
    mission to, 112-115, 116, 117

V

Valhalla Basin on Callisto, *66,* 67
Van Allen, James, 54, 84
Volcanism, Io's, *2-3,* 61, *62-63, 64,*
    65-66
Voyager spacecraft, 37, 58-60;
    high-energy telescope, *41;* images
    from, *42-43, 59,* 60, *61-64,* 65,
    66-67, 75, 85, 86, *88, 108-109, 114,*
    116, *118-119;* imaging problems of,
    handling, 90, 111-112, 113;
    Mariner/Jupiter-Saturn program
    and, 57-58; Saturn missions of,
    85-90; *Voyager 1* pictures, *42-43,*
    60, *62-63,* 65, *75, 85,* 86; *Voyager 2,*
    37, *39,* 58-60, 61, 64, 81, 88-90, 109,
    110, 111-115, 116, 117, 118, 120

W

Warwick, James, 55, 115
Watson, William, 20; Herschel's
    letter to, quoted, 19
Waves, spiral density, and Saturn's
    rings, 98, *102-103*
Weather: Jupiter's, *74-75;* Saturn's,
    89, 91-92; Uranus's, 114-115
Wildt, Rupert, 46-48
Winds: of Jupiter, *74-75;* of Saturn,
    92; solar, *40,* 54, 76; of Uranus,
    114-115
Wolfe, John, quoted, 84

Y

Young, Tom, 80-81; quoted, 84

ACKNOWLEDGMENTS

The editors of *The Far Planets* also wish to thank these people for their contributions: Arthur Adel, Northern Arizona University, Flagstaff; Richard P. Binzel, Planetary Science Institute, Tucson, Ariz.; Bill Booth, The Franklin Institute, Philadelphia; Marc Buie, University of Hawaii at Manoa, Honolulu; Giorgio Buonvino, Osservatorio Astronomico, Rome; Bernard F. Burke, Massachusetts Institute of Technology, Cambridge; John E. P. Connerney, NASA Goddard Space Flight Center, Greenbelt, Md.; Frank Drake, University of California at Santa Cruz; James Elliot, Massachusetts Institute of Technology, Cambridge; Fraser Fanale, University of Hawaii at Manoa, Honolulu; Louis A. Frank, University of Iowa, Iowa City; Kenneth L. Franklin, McAllen, Tex.; Gerhard Hartel, Rainer Herbster, Deutsches Museum, Munich; Peter Hingley, Librarian, Royal Astronomical Society, London; Andrew Ingersoll, California Institute of Technology, Pasadena; Torrence Johnson, Jet Propulsion Laboratory, Pasadena, Calif.; Zeeann Mason, The Franklin Institute, Philadelphia; Mara Miniati, Istituto e Museo di Storia della Scienza, Florence; Giuseppe Monaco, Curator, Museo Astronomico e Copernicano, Rome; Derrick Pitts, The Franklin Institute, Philadelphia; Eugene Shoemaker, U.S. Geological Survey, Flagstaff, Ariz.; Jurrie J. van der Woude, Jet Propulsion Laboratory, Pasadena, Calif.; Cliff Wagner, The Franklin Institute, Philadelphia; James Warwick, University of Colorado, Boulder; Gerd Weigelt, Physikalisches Institut, Erlanger-Nurnberg, West Germany; Robert A. West, Jet Propulsion Laboratory, Pasadena, Calif.; Jim Young, Jet Propulsion Laboratory/Table Mountain Facility, Wrightwood, Calif.

PICTURE CREDITS

The sources for the illustrations that appear in this book are listed below. Credits from left to right are separated by semicolons, from top to bottom by dashes.

Cover: Painted by Kazuaki Iwasaki. Front and back endpapers: Art by John Drummond. 2-7: Art by Paul Hudson. 12, 13: Art by Damon Hertig. 14: Computer-generated initial cap by John Drummond, detail from photo appearing on pages 12, 13. 15: Art by John Drummond. 16: Art by Damon Hertig. 18, 19: Donato Pineider, courtesy Biblioteca Nazionale Centrale, Florence. 22, 23: Art by Yvonne Gensurowsky of Stansbury, Ronsaville, Wood Inc. 24, 25: Art by Stansbury, Ronsaville, Wood Inc. 26-29: Art by Damon Hertig. 30, 31: Art by Fred Holz—art by Damon Hertig. 33: Art by Stansbury, Ronsaville, Wood Inc. 34, 35: Lowell Observatory Photographs. 38, 39: Art by Dennis Davidson. 40, 41: Art by Fred Holz. 42, 43: JPL/NASA. 44, 45: NASA/JPL. 46: Computer-generated initial cap by John Drummond, detail from photo appearing on pages 44, 45. 48, 49: Art by Damon Hertig. 50, 51: Art by Stansbury, Ronsaville, Wood Inc. 53: NASA/JPL-Table Mountain Observatory, California. 54: Bob Brown, Space Telescope Science Institute; the NASA Infrared Telescope Facility and the University of Hawaii in connection with NASA. Line art by John Drummond. 55: Courtesy NRAO/AUI. Line art by John Drummond. 59: NASA/JPL. 61: JPL/NASA—art by Fred Holz. 62, 63: Art by Stansbury, Ronsaville, Wood Inc.; NASA, Washington, D.C.; USGS, Flagstaff, Arizona (2). 64: NASA/JPL. 66, 67: JPL. 68: Art by Damon Hertig. 70, 71: Art by Stansbury, Ronsaville, Wood Inc. 72, 73: Art by Stansbury, Ronsaville, Wood Inc.—JPL/NASA. 74: Art by Stansbury, Ronsaville, Wood Inc. 75: JPL/NASA. 76, 77: Art by Stansbury, Ronsaville, Wood Inc.; art by John Drummond. 78, 79: Anglo-Australian Observatory. 80: Computer-generated initial cap by John Drummond, detail from photo appearing on pages 78, 79. 83: NASA. 85: NASA/JPL. 87: Art by John Drummond from drawing courtesy the British Library, London. 88: NASA/JPL—art by John Drummond. 91: Courtesy NRAO/AUI. 93, 94: Art by Stansbury, Ronsaville, Wood Inc. 95: Art by Stansbury, Ronsaville, Wood Inc.; art by Yvonne Gensurowsky of Stansbury, Ronsaville, Wood Inc. 96: NASA/JPL. 99: Art by Fred Holz. 100, 101: Art by Joe Bergeron. 102, 103: Art by Joe Bergeron; art by John Drummond. 104, 105: Art by Joe Bergeron. 106, 107: Art by John Drummond; art by Joe Bergeron. 108, 109: NASA, Washington, D.C. 110: Computer-generated initial cap by John Drummond, detail from photo appearing on pages 108, 109. 114-117: NASA/JPL. 118: Richard Berry *Astronomy* magazine from JPL; NASA/JPL. 119: NASA, Washington, D.C. 120, 121: JPL/NASA. 122: U.S. Naval Observatory—G. Baier and G. Weigelt, Physikalisches Institut, Erlangen, West Germany; University of Arizona, Steward Observatory. 123-127: Art by Douglas R. Chezem. 128: Art by Damon Hertig. 130-133: Art by Paul Hudson. 140, 141: Art by John Drummond.

**Library of Congress Cataloging in
Publication Data**
The far planets/by the editors of Time-Life
Books.
p.   cm.   (Voyage through the universe).
Bibliography: p.
Includes index.
ISBN 0-8094-6854-9.
ISBN 0-8094-6855-7 (lib. bdg.).
1. Planets. I. Time-Life Books. II. Series.
QB601.F37 1989
523.4—dc19   88-15994 CIP

For information on and a full description of any
of the Time-Life Books series, please call 1-800-
621-7026 or write:
Reader Information
Time-Life Customer Service
P.O. Box C-32068
Richmond, Virginia 23261-2068